NATURAL

SOAP

NATURAL SOAP

MELINDA COSS

PLAIN SIGHT PUBLISHING
AN IMPRINT OF CEDAR FORT, INC.
SPRINGVILLE, UTAH

The formulations in this book are copyrighted and cannot be made for resale.

The soap recipes in this book are not directly converted from metric to imperial based on 28.35 grams per ounce but on 25 grams per ounce. This adjustment was made to the recipes to achieve a safe and easy-to-weigh alternative. Please use either metric or imperial measurements, but don't mix the two.

The information in this book has been carefully researched and all efforts have been made to ensure accuracy. The author and publisher accept no responsibility for any injuries, damage, or losses incurred either during, or subsequent to, following the instructions in this book.

ISBN 13: 978-1-4621-1242-5

Published by Plain Sight Publishing, an imprint of Cedar Fort, Inc.
2373 W. 700 S. Springville, UT 84663
Distributed by Cedar Fort, Inc., www.cedarfort.com

LIBRARY OF CONGRESS CATALOGING-IN-PUBLICATION DATA

Coss, Melinda, 1949-
Natural soap : techniques and recipes for beautiful handcrafted soaps, lotions, and balms / Melinda Coss.
  pages cm
Originally published: United Kingdom: New Holland Publishers, 2011.
Includes bibliographical references and index.
Summary: Instructions on how to make soaps, lotions, and balms.
ISBN 978-1-4621-1242-5 (alk. paper)
1. Soap--Handbooks, manuals, etc. I. Title.

TP991.C77 2013
668.'12--dc23

2012050708

Cover design and typesetting by Angela D. Olsen
Cover design © 2013 by Lyle Mortimer
Edited by Whitney A. Lindsley

Printed in China

10 9 8 7 6 5 4 3 2 1

Contents

Introduction

THE VAST MAJORITY OF MY WORKING life has been spent in the design and marketing of craft products. For many years, I was involved in knitting and embroidery and wrote numerous books on those subjects, but in the mid-nineties, I discovered soapmaking, and my life was turned completely upside down. It is hard to imagine how a simple bar of soap can fundamentally change lives, but history shows that it truly can.

First, if you look at soap as a simple hygiene product, it has and does undoubtedly fight disease and save lives. In addition, soap presents creative marketing people with an irresistible blank canvas. They begin with a product that is used by men, women, and children worldwide in all their wonderful shapes, colors, sizes, and income brackets, and create, design, and package a simple bar of soap to fulfill needs, lifestyles, and aspirations. I am frequently asked the question "can you really make a living making and selling soap?" The answer is that 70 percent of people use soap and the other 30 percent just need to be persuaded. This is a dream market for an entrepreneur, but your success or failure will depend on your business acumen, your creativity, your marketing skills, and your product, in that order.

Soapmaking sent me on the hugely challenging adventure of building an enterprise from scratch with traditional values and methods and with the aim of making enough volume to supply supermarkets. My creative and business learning curve was steep and fruitful, and in the last five years, I have had the privilege of sharing that learning curve with entrepreneurial spirits worldwide through my consultancy work and my teaching hands-on soapmaking to the many hundreds of people who have attended my courses.

An even greater privilege has come from the opportunity to start social soapmaking enterprises in Africa and to discover and add value to local plants and oils that can be used in soaps and toiletries. This work has been humbling and hugely satisfying.

If you have my two earlier books, *The Handmade Soap Book* and *Gourmet Soaps Made Easy*, you will note many differences in method in my bar soap section. These changes come from years of experimentation and practical experience. My previous methods still work, but the processes suggested in this book will make your life a lot easier.

I am also including sections on liquid soapmaking and on creams and lotions. Liquid soap outsells bar soap eight to one, so if you want to run a business, it's a good skill to master. Creams and lotions give you the opportunity of experimenting with all the wonderful luxury oils without worrying about caustic substances—the products you can produce at home in your kitchen are wonderful, but don't take my word for it; make them and see for yourself!

Finally, for those soap entrepreneurs of the future, I have included a section on setting up a business.

Enjoy!

Ingredients

SOAP IS THE RESULT OF COMBINING AN alkali with an acid. The alkali used to make bar soap is sodium hydroxide (also known as caustic soda, NaOH, and lye), and the alkali used to make liquid and soft soaps is potassium hydroxide. The acid can be any kind of oil or fat, be it animal or vegetable. If you think of all the wonderful plant-derived oils and fats that are available, you will realize what a huge palette this gives you to play with. In the following section, I will explain the virtues of various oils and additives and hopefully help you to make your ideal product.

Creams and lotions are a combination of oils/fats/butters, water additives, and an emulsifier to hold them together. They contain no harsh chemicals, and making them is a similar experience to that of making mayonnaise. Reserve the use of your finest grade base and essential oils for making creams and lotions—here the benefits will really come through.

Once you have your basic product formula, there are many ingredients you can use to add color, texture, and fragrance, and this section should give you plenty of ideas.

If you think sensibly about the fact that oils are mixed with a caustic solution to create soap, you must also realize that the caustic element will, certainly to some degree, destroy the natural benefits of any chosen oil. For this reason, and for reasons of economy (oils are becoming costly), it is wise not to invest in top-grade refined oils when making soap. Many companies boast the use of "virgin olive oil" in their products, but in reality "pomace" (the roughest grade of olive oil produced from the third pressing of the fruit) creates far better soap. Some olive oils sold as "pomace" are in fact blended, so take care to read labels before you buy since you need a 100 percent olive pomace oil. Essential oils need not be top grade, but they should be 100 percent pure and not blended with alcohol or enhancers. Always buy your ingredients from a reputable supplier (the stockist list on page 143 should help you to find one).

CHOOSE WHAT YOU USE

Sales of natural and pseudo-natural soaps and cosmetics are rising year by year. There are several sound reasons for this, the first being that we live in a world where our tolerance to chemical household products is diminishing, and there is a sharp increase in the number of people suffering from skin allergies. The second reason is that any right-thinking person wants to do their bit for the environment and to leave a sustainable world behind them for future generations to enjoy. The third reason is a growing awareness that fair-trade sales of natural botanical products help to create and sustain agricultural industries, thus alleviating poverty in underdeveloped countries. What a wonderful product we have here!

In order to meet the growing demand for natural products, many manufacturers choose to express "naturalness" in their brand names and include minute amounts of botanical oils. The way to beat this trend is to read the label on the package and to be aware of certain ingredients that are widely in use but are in fact proven to be skin irritants or, even worse, carcinogenic. That said, I do not support the argument that "natural" is without exception "good"—we need to make informed decisions on what we buy. Remember that arsenic (as an extreme example) is 100 percent natural!

If you suffer from eczema or sensitive skin, you should not use products that contain any colorings or fragrance—natural or unnatural—as these are the biggest contributors to skin irritation. Some essential oils (such as Roman chamomile) will help mildly sensitive skin and are safe to use in baby products, but only in tiny quantities. Some soapmakers, aware that "no smell means no sell," bombard their products with large quantities of essential oils, and while the soap may smell divine, doing so is downright dangerous. In the European Union (EU), cosmetic legislation dictates that a maximum of 2 percent essential oils can legally be used in soap products, but in fact 1 to 1.5 percent is a much safer amount to aim for.

When reading the labels on your storebought soaps and cosmetics, be aware of the following:

Sodium lauryl sulphate (SLS)
This is a detergent derived from coconut oil and is widely used in shampoos. It contains endocrine disruptors and estrogen mimics, and it can damage the skin barrier functions, which will increase the allergic response to other toxins and allergens.

Sodium laureth sulfate (SLES)
Found in shampoo, toothpaste, bubble bath, body wash, and soap, this is another known endocrine disruptor and estrogen mimic, and it is also carcinogenic. It allows other chemicals to penetrate skin more deeply and potentially enter the bloodstream.

Mineral oil and petroleum jelly
These petroleum-based ingredients block pores, act as barriers, and suppress normal skin functions.

Parabens (methylparaben, propylparaben, ethylparaben, butylparaben)
These are used to preserve cosmetic products. There is a growing public awareness about the dangers of using parabens, and they have been linked to breast cancer. The counterargument is that parabens aren't dangerous at low levels, and they are stable preservatives, protecting products against far more dangerous viruses. On the market are a number of safer preservatives that can be used in your natural products. However, despite the regular industry proclamations of the discovery of a wholly natural preservative (such as GSE, ROE, and citricidal), none of these products will pass a challenge test.

Borax (sodium borate)
Widely used in the United States to stabilize and thicken liquid soap, Borax is restricted under EU cosmetic legislation because it is a strong irritant. It can cause rashes in babies and young children.

DEA (diethanolamine, cocamide DEA, oleamide DEA, and lauramide DEA)
These ingredients are used to increase lather in industrially produced soap and have been known to react with other ingredients to create a carcinogen that can easily be absorbed by the skin.

Even when using entirely natural oils, it is important that you establish where they were grown, and for ecological reasons, avoid anything produced in an unsustainable environment. Palm oil is a particular case in question. Palm oil is used in one out of every ten products on the supermarket shelves and widely used in biofuel, so the rush to create palm plantations, particularly in the tropical rainforests of Malaysia and Indonesia, has led to the destruction of forests that support endangered wildlife, particularly tree-dwelling primates. Palm oil is traditionally a soapmaking staple, and you can buy it from sustainable sources—just make sure you ask your potential supplier the right questions.

OILS, FATS, AND BUTTERS
The following list provides the International Nomenclature of Cosmetic Ingredients (INCI) name of a variety of oils, fats, and butters that can be used in your products, and describes the principal properties To help you understand the terminology, the table on page II shows what the fatty acids actually contribute to your soaps in terms of quality and feel, and the higher the percentage specified next to the oil, the greater the contribution. The iodine value also contributes to the hardness of a soap bar, but in this case, the lower the value, the harder the bar.

Oils marked with an asterisk (*) are basic, relatively affordable soapmaking oils otherwise known as "base oils." The others are quite valuable but can be used in small quantities in your soaps. The really expensive oils should be reserved for your cream and lotion recipes.

If an oil is described as being high in antioxidants, this means it is full of vitamins, minerals, and enzymes that will protect both your skin and the product itself from degeneration.

FATTY ACID	PROPERTIES IT LENDS TO SOAP
LAURIC ACID	Hard bar, cleansing, fluffy lather, fast trace
LINOLEIC ACID	Conditioning
MYRISTIC ACID	Hard bar, cleansing, fluffy lather
OLEIC ACID	Conditioning
PALMITIC ACID	Hard bar, stable lather
RICINOLEIC ACID	Conditioning, fluffy stable lather
STEARIC ACID	Hard bar, stable lather

ALMOND OIL (sweet) *

Good base for creams and body oils, high in Omega 9 and vitamins A and B, makes a white bar of soap. Good, inexpensive base for creams and lotions.

INCI: *Prunus dulcis*
Oleic: 64–82%
Linoleic: 8–28%
Palmitic: 6–8%
Iodine: 93–106

APRICOT KERNEL OIL

Good for sensitive, dry, and problem skins; has excellent skin-softening qualities.

INCI: *Prunus armeniaca*
Oleic: 58–74%
Linoleic: 20–34%
Palmitic: 4–7%
Iodine: 92–108

ARGAN NUT OIL

Rich in antioxidants and vitamin E, an excellent choice for mature or dry skins.

INCI: *Argania spinosa*
Oleic: 42.8%
Linoleic: 36.8%
Palmitic: 12%
Stearic: 6%
Iodine: 98

AVOCADO OIL *

A semi-fatty oil with a high vitamin content. A good choice for problem skins.

INCI: *Persea gratissima*
Oleic: 36–80%
Palmitic: 7–32%
Linoleic: 6–18%
Stearic: 1.5%
Iodine: 82–90

BABASSU OIL *

Properties similar to those in coconut oil with high lauric acid content. Rich and moisturizing.

INCI: *Orbygnia oleifera*
Lauric: 50%

Myristic: 20%
Palmitic: 11%
Oleic: 10%
Stearic: 3.5%
Iodine: 10–20

BAOBAB OIL *

Rich in Omega acids and vitamin D. Creates a barrier that protects the skin.

INCI: *Adansonia digitata*
Palmitic: 24%
Oleic: 36%
Linoleic: 31%
Iodine: 88

BORAGE OIL

High in gama linolenic acid, helps to lock in moisture and smooth the skin. An ideal choice for a baby cream.

INCI: *Borago officinalis*
Palmitic: 10–11%
Oleic: 16–20%
Linoleic: 35–38%
Stearic: 3.5–4.5%
Iodine: 165–185

BLACK CUMIN OIL

Rich in essential fatty acids, good for problem skin, and promotes smoothness.

INCI: *Nigella sativa*
Oleic: 64%
Palmitic: 11.7%
Linoleic: 12%
Linolenic: 70%
Iodine: 110–125

CALENDULA (MARIGOLD) OIL

Antiseptic and anti-inflammatory, excellent oil for damaged and sensitive skin.

INCI: *Calendula officinalis*

Calendula oil is made by macerating the flowers in a base oil. Therefore the fatty acid values relate to the specific base oil used, rather than the calendula itself. Check the identity of the base oil with your supplier.

CANOLA OIL *

Used mainly as an inexpensive filler in soapmaking, canola or rapeseed oil does tend to produce a soft, low-lathering soap and is best used in combination with an equal or higher percentage of coconut oil.

INCI: *Canola* for canola oil or *Brassica campestris* for rapeseed oil

Oleic: 32%
Linoleic: 15%
Palmitic: 1%
Iodine: 105–120

CASTOR SEED OIL *

Used in soap for its high-foaming properties, this oil is particularly useful in shampoos and shaving soaps but does accelerate trace.

INCI: *Ricinus communis*

Ricinoleic: 90%
Linoleic: 3–4%
Oleic: 3–4%
Iodine: 82–90

COCOA BUTTER

High in saturated fats and vitamin E, cocoa butter is a natural moisturizer that will add hardness to your soap and body to your creams.

INCI: *Theobroma cacao*

Oleic: 34–36%
Stearic: 31–35%
Palmitic: 25–30%
Linoleic: 3%
Iodine: 33–42

COCONUT OIL *

Copra and RBD (refined, bleached, and deodorized) coconut oils are the soapmaker's staple, creating a hard bar soap with a rich foam. Used alone, it can be drying, so it should be combined with other oils. The rarer virgin coconut oil is high in saturated fatty acids and acts as a good moisturizer and thickener in creams.

INCI: *Cocos nucifera*

Lauric: 39–54%
Myristic: 15–23%
Palmitic: 6–11%
Oleic: 4–11%
Stearic: 1–4%
Linoleic: 1–2%
Iodine: <10

COMFREY OIL

Macerated in sunflower or olive oil, comfrey contains allantoin, a cell-proliferant that helps repair damaged tissue.

INCI: *Symphytum officianale*

The fatty acid profile for comfrey oil depends on the base oil it is macerated in. Clarify this with your supplier.

EMU OIL

This is an anti-inflammatory oil popular with Australian and South African soapmakers, and is becoming more so in the United States. While I personally can't bring myself to try soap made from emus, reports are that it adds greatly to the moisturizing qualities of bar soap.

INCI: Emu oil

Linoleic: 14%
Myristic: 0.4%
Oleic: 50%
Palmitic: 21%
Stearic: 9%
Iodine: 75

EVENING PRIMROSE OIL

High in gamma-linolenic acid, excellent in creams and lotions for mature skin.

INCI: *Oenothera biennis*

Palmitic: 5.5–7%

Oleic: 5–11%

Stearic: 1.5–2.5%

Linoleic: 70–77%

Iodine: 135–165

GRAPESEED OIL

A light base oil that is easily absorbed by the skin, firms up, and regenerates. Rich in vitamins and minerals.

INCI: *Vitis vinifera*

Linoleic: 58–78%

Oleic: 12–28%

Palmitic: 5–11%

Stearic: 3–6%

Iodine: 125–142

HEMPSEED OIL *

High percentage of essential fatty acids and vitamins, moisturizing, and helps to balance dry skin. Makes a gorgeous soap but used on its own has a short shelf life.

INCI: *Cannabis sativa*

Oleic: 12%

Linoleic: 57%

Palmitic: 6%

Stearic: 2%

Linolenic: 21%

Iodine: 166.5

JOJOBA

Actually a wax, not an oil, jojoba easily penetrates the skin and provides a protective barrier. Adds thickness to creams and is very conditioning when used in hair products.

INCI: *Simmondsia chinensis*

Oleic: 10–13%

Iodine: 80–85

LARD (pig fat) *

Makes a hard, cleansing soap bar with plenty of foam.

INCI in soap: *Sodium lardate*

Oleic: 46%

Palmitic: 28%

Stearic: 13%

Linoleic: 6%

Myristic: 1%

Iodine: 43–45

MACADAMIA NUT OIL

High in omega-9, this fatty oil penetrates and moisturizes dry and mature skin.

INCI: *Macademia tetraphylla*

Oleic: 54–63%

Palmitic: 7–10%

Stearic: 2–6%

Linoleic: 1–3%

Iodine: 73–79

MARULA OIL

Great antioxidant, originally used on its own as a cleansing, moisturizing body oil.

INCI: *Sclerocarya birrea*

Oleic: 70–78%

Palmitic: 9–12%

Stearic: 5–8%

Linoleic: 4–7%

Iodine: 70–76

MEADOWFOAM OIL

A stable oil that is great for dry or sensitive skin. Highly moisturizing and rejuvenating, it is great in creams and lotions.

INCI: *Limnanthes alba*

Oleic: 3.2%

Iodine: 92

MELON OIL (Kalahari)

Rich in fatty acids and vitamin C, detoxifying and rejuvenating. Light and easily absorbed by the skin.

INCI: *Citrallus lanatus*

Oleic: 16–17%

Palmitic: 11–12%

Stearic: 7.5–8%

Linoleic: 50–70%

Iodine: 120–130

OLIVE OIL *

The first choice of many soapmakers, pure olive oil soaps get better as they get older but trace and harden very slowly. You either love or hate the slightly oily feel and the slow lather. A relatively low-cost option for a cream or lotion base.

INCI: *Olea europaea*

Oleic: 63–81%

Palmitic: 7–14%

Linoleic: 5–15%

Stearic: 3–5%

Iodine: 79–95

PALM OIL *

Provided this is purchased from an ecologically-sound source, it is a staple oil for soapmakers providing an inexpensive, hard, long-lasting soap bar with a relatively low foam. Palm oil is produced from the fleshy fruit of the oil palm.

INCI in soap: *Elaeis guineensis*

Palmitic: 43–45%

Oleic: 38–40%

Linoleic: 9–11%

Stearic: 4–5%

Iodine: 45–57

PALM KERNEL OIL *

Widely used as a base in industrial soap production, this is a good bulking oil and has a different fatty acid makeup from palm oil. Because of the high lauric content, this oil traces very quickly. Palm kernel oil is produced from the kernel of the oil palm fruit.

INCI: *Elaeis guineensis*

Lauric: 47%

Oleic: 18%

Myristic: 14%

Palmitic: 9%

Iodine: 37

PEACH KERNEL OIL

High in protein, vitamins A and E, and omega-6.
Good for all skin types and lovely in creams.
INCI: *Prunus persica*
Oleic: 55–75%
Palmitic: 5–8%
Linoleic: 15–35%
Iodine: 108–118

ROSEHIP SEED OIL

Rich in vitamins A and C, this cell regenerator is
great for stretch marks and scar tissue.
INCI: *Rosa mosqueta*
Palmitic: 3–5%
Oleic: 10–20%
Linoleic: 26–37%
Stearic: 1–3%
Iodine: 179–192

SEA BUCKTHORN OIL

High in vitamin C and B-carotene, this natural
antioxidant is known to combat wrinkles,
eczema, and dry skin.
INCI: *Hippophae rhamnoides*
Palmitic: 34–35%
Linoleic: 35–36%
Iodine: 86

SUNFLOWER OIL *

High in fatty acids and easily absorbed by the
skin, a good base oil for all skin types. Great
option for liquid soap and as a bulk oil in bar
soaps, although it should be combined with
coconut oil, or the soap will be very soft.
INCI: *Helianthus anuus*
Linoleic: 70%
Oleic: 16%
Palmitic: 7%
Stearic: 4%
Iodine: 119–138

SHEA BUTTER (KARITE)

High in saturated fats, an effective moisturizer
that stimulates cell growth, good for stretch marks
and scar tissue. Great moisturizer and hardener
when used in bar soaps. Thickens creams.

INCI: *Butyrospermum parkii*
Oleic: 40–55%
Stearic: 35–45%
Linoleic: 3–8%
Palmitic: 3–7%
Iodine: 55–71

TALLOW (beef fat) *

A traditional soapmaking ingredient that
produces a hard, conditioning soap bar.
INCI in soap: *Sodium tallowate*
Oleic: 37–43%
Palmitic: 24–32%
Stearic: 20–25%
Myristic: 3–6%
Linoleic: 2–3%
Iodine: 43–45

WHEATGERM OIL

An antibacterial that stimulates skin cells. High
in vitamins and essential fatty acids but goes
rancid quickly.
INCI: *Triticum vulgare*
Linoleic: 55–60%
Oleic: 13–21%
Palmitic: 13–20%
Stearic: 2%
Iodine: 125–135

VITAMIN E OIL

A natural antioxidant that will extend the life of
your products. Smooths dry and rough skin.
INCI: Tocopheryl

SCENTS AND SENSIBILITIES

One of the most important elements of your soap is fragrance. Those of us who want our products to be totally natural opt for essential oils diffused or pressed from plants or citrus fruits, used both for their fragrance and for their therapeutic value. The second option is synthetic fragrance oils. These can imitate scents that would be too expensive to achieve naturally (such as rose), and they can also mimic fruit and even lemon meringue pie if that is the perfume you are trying to achieve. Whatever your preference, the one important fact you need to bear in mind if you are planning on marketing your soap is "no smell, no sell."

In mass-produced soap, fragrance is added well after the chemical reaction is complete, and it is therefore stable and strong in the soap. In cold processed soapmaking, we add the fragrance while the chemical reaction is going on, and, therefore, we have to be careful that the blend of oils we choose will withstand this saponification process (conversion to soap). The extensive art of aromatherapy would fill this book all on its own, but basically, essential oils are split into three categories or "notes," being top, middle, and base notes. A "top note" fragrance is the first to reach your nose but can be fleeting, a "middle note" is the core of the fragrance blend, while a "base note" anchors the blend and gives it a sensuous undertone. When composing a blend for a cold process soap, it helps considerably if you use a combination of the three notes, or at least include a base note oil.

Another useful trick is to add your essential oil to dry matter before you put it in your soap. If you plan to put herbs, petals, or clays in your soap, pour the essential oils over them first—this process helps to suspend the fragrance in the soap mix and increases its holding power. When working with liquid soap, you don't have this problem because the saponification is complete before you add the fragrance.

When buying essential oils, always go to a reputable supplier and ensure that the oils are 100 percent pure. This does not mean you have to

use the most expensive varieties, but do check that oils do not contain dilutants because these may interfere with the saponification process and cause your soap mixture to seize. Likewise, when using artificial fragrances, only buy from a supplier who understands cold process soapmaking and make sure the fragrances are of a cosmetic grade and not just produced for fragrancing a room or for candle making.

If you are making toiletries for yourself, perhaps because you have allergies or sensitive skin, you must bear in mind that both natural and unnatural fragrances can be high in allergens, and these, along with pigments, are the biggest cause of skin irritation. In Europe, by law, your fragrance content must not be more than 2 percent of your total soap formula, and this is for good reason. Essential oils are extremely concentrated and powerful and, while when used with knowledge and care they are undoubtedly beneficial to the skin, if you use them heavily just to increase fragrance, they can be dangerous. European Union cosmetic legislation now also requires you to list on your product labels some specific allergens that occur naturally within many essential oils.

When making creams and lotions, you can create a light scent by replacing the water content with a flower water or hydrosol (see page 125). Because only a tiny amount of essential oil is needed to fragrance a cream, you can also use some of the precious essential oils such as neroli or Roman chamomile, both of which are reasonably safe for those prone to eczema or dermatitis. Below are some of the more popular essential oils and their properties. You will also find a chart on pages 110–111 that will help you choose fragrances for specific skin types.

CARDAMOM (*Elletaria cardamomum*)
TOP NOTE Extracted from the plant seed, this pungent and spicy oil works well with blends of cinnamon, orange, and frankincense. Irritant, so use only in small quantities.

CEDARWOOD ATLAS (*Cedrus atlantica*)
BASE NOTE Distilled from the wood, this refreshing, clean fragrance is great for masculine toiletries and holds up well in cold process soap.

CHAMOMILE (Roman) (*Athemis nobilis*)
TOP NOTE Distilled from the flowers and leaves, a great choice for those with sensitive skin, including babies.

CINNAMON LEAF (*Cinnamomum verum J.*)
MIDDLE NOTE Not to be confused with cinnamon bark oil (which should not be used), cinnamon leaf oil is fine in small quantities and blends beautifully with ginger and orange.

CLOVE (*Eugenia caryophyllata—bud*)
BASE/MIDDLE NOTE Useful boost for spicy blends but should be used in tiny quantities since it is high in allergens. Should not be used during pregnancy. Accelerates trace in soaps.

EUCALYPTUS (*Eucalyptus globulus*)
TOP NOTE Powerful clean scent, good for blending with citrus fragrances, antifungal and antiseptic.

FRANKINCENSE (*Boswellia carterii*)
BASE NOTE Distilled gum resin and precious oil with a deep, spicy, sweet fragrance. Blend with lavender and neroli for a real treat.

GERANIUM (*Pelargonium graveolens*)
MIDDLE NOTE The Egyptian variety is reasonably priced and holds well in soap. This is the closest you will get to a rose scent without breaking the bank. Delicious heady, floral fragrance.

GINGER (*Zingiber officinale*) MIDDLE NOTE Adds an interesting twist to citrus blends but can cause irritation so only use in small amounts.

GRAPEFRUIT (*Citrus paradisi*)
TOP NOTE Does not hold well in cold process soaps but a good choice for liquid soap. Nice addition to lavender.

HOWOOD LEAF (*Cinnamomum camphora*)
MIDDLE NOTE Only use in small quantities due to the possibility of skin irritation. Has a sweet, earthy fragrance that blends well with sweet orange and lavender.

LAVENDER (*Lavendula officinalis*) TOP NOTE A wonderful staple oil that holds up well. Even if you hate lavender, understand that lavender soap outsells all other varieties eight to one.

LEMON (*Citrus limonum*) TOP NOTE Therapeutic and healing when used in creams, but unless cleverly blended does not hold up well in cold process soap. Try *Litsea cubeba* or lemongrass as a refreshing citrus alternative.

LEMONGRASS (*Cymbopogon citratus*)
TOP NOTE Refreshing citrus fragrance that holds well in cold processed soap and blends well with lavender. Excellent choice for men but can cause irritation, so keep the percentages low.

MANDARIN (*Citrus reticulate*) TOP NOTE A citrus scent with floral undertones. Lovely in liquid soap and creams, but no staying power in cold processed soaps.

MAY CHANG (*Litsea cubeba*) TOP NOTE Has a great citrus scent that can be used to anchor other scents. Holds up well in cold process soap.

NEROLI (*Melaleuca viridiflora*) MIDDLE NOTE Used as the basis for eau de cologne, this bitter orange blossom scent is highly prized and priced so should not be wasted in cold processed soap.

PALMAROSA (*Cymbopogon martini*)
MIDDLE NOTE Fresh, sweet, and floral with
a rose geranium overtone. Has moisturizing
and balancing properties.

PATCHOULI (*Pogostemon patchouli*)
BASE NOTE The stuff of incense sticks and
cachous, patchouli is a wonderful anchor for
blends and brings on waves of 1960s nostalgia
when used on its own. Lovely with geranium
and/or lavender.

PEPPERMINT (*Metha piperata*)
TOP NOTE A stimulating oil that is good in foot
creams and insect repellents. Should not be used
during the first three months of pregnancy.

TEA TREE (*Melaleuca alternifolia*)
TOP NOTE Antiseptic in both purpose and
fragrance and will help to preserve creams and
lotions. Good for insect bites and acne.

Ylang-ylang (*Cananga odorata*) MIDDLE NOTE
Heady floral fragrance that holds up well in cold
process soap and adds a touch of romance to a
blend of oils.

HERBS, SPICES, AND ADDITIVES

Once you have decided on the oils you want to
use in your base recipe, there are numerous
delights you can add at the trace stage (when
your soap thickens) to enhance the texture and
performance of your product. This is the fun
part of soapmaking.

If you want a bar to have exfoliating qualities,
experiment by adding dried herbs, poppy seeds,
pumice powder, ground orange or lemon peel,
dried seaweed, oatmeal, bran (or other pulses),
salt, or sugar. There are also a vast number of
seeds and fruit powders on the market, such
as raspberry and strawberry seeds and ground
fruit kernels (try olive or apricot for a scrubby
effect).

Dried flower petals (rose, lavender, and
calendula) can also be added at trace, but bear in
mind that only calendula (marigold) will retain
its color—sadly, everything else turns brown over
time.

The amount of dried additive you use is
purely a matter of taste and has no bearing on
the saponification process, but if you are using
any dried "live matter," it is a good idea to add
1 percent vitamin E oil to your formula to help
preserve it.

Pureed fruit or aloe vera can also be added to
your soap. Deduct the weight from the weight of
the water in your formula and add the pulp/gel
to the lye water before mixing the soap.

Argile clay, added at trace, gives your soap "slip"
and helps to draw out any impurities in the skin.
It can also be used in creams and is available in
green, pink, yellow, and red, so it adds natural
color (see page 59). Coffee grounds act as
deodorizers and are a good option for kitchen
soap since they will help to get the smell of garlic
and onions off your hands. Simply make a pot
of coffee, drain out the coffee grounds, and add
these at trace. If you want a dark brown soap, use
actual coffee instead of water to make up your lye.

For a harder soap, try adding up to 2 percent
beeswax to your formula. Beeswax is also a useful
emulsifier and thickener for creams and is

particularly good in pure olive oil soaps. Honey is a scrumptious additive and is used for its emollient qualities. It can be added at trace if warmed slightly before use, or alternatively add it to your lye solution.

COLORINGS

The earth provides us with a number of natural ingredients that can be used to color soaps and creams. The following natural ingredients are particularly useful for muted colors. Dilute the colorings in a drop of warm water and then mix with a spoonful or two of your traced soap before adding this back to your main batch and mixing thoroughly. This will avoid a speckled effect.

CARROT POWDER	ORANGE
CAYENNE PEPPER	SALMON
COCOA POWDER	BROWN
COFFEE	BROWN
CURRY POWDER	YELLOW TO PEACH
GINGER POWDER	PEACH
MADDER	PINK
PAPRIKA	PEACH
SPIRULINA	GREEN
TURMERIC	CORAL

Two useful gifts from nature are alkanet, which is a tree bark, and the seed annatto, commonly used to color cheese. Both of these should be infused in warm sunflower or olive oil before use. Simply heat a cupful of oil and add ⅓ of a cupful of bark or seeds and leave this to infuse until it is deep purple, or yellow in the case of annatto. Alkanet will turn your soap anything from gray to a rosy lavender depending on the alkalinity of the soap while annatto gives you a warm sunshine yellow.

If you want to achieve bright colors there is no choice but to use cosmetic grade oxide powders or FD&C or D&C (food drug and cosmetic) grade manmade pigments. Oxides are sometimes sold as "natural" but in reality they are an inorganic compound made up from synthetically prepared iron oxides that include some hydrated form of naturally occurring mineral deposits.

FD&C and D&C colorants are manmade, but tiny amounts are needed to color your products. For this reason, they are considered entirely safe.

When using oxides, dissolve approximately ¾ tsp in 1 oz (28 g) of warm distilled water—this is sufficient for a 2 lb (1 kg) batch of soap. When using FD&C and D&C, colorants you need only enough pigment to cover the tip of a teaspoon dissolved in the same amount of water to produce quite a strong color. Add these colors drop by drop at trace until you achieve the shade you want. For liquid soap and creams, you can use liquid food colorants, but be aware that only one drop of colorant will give you quite a strong color.

Bar Soaps

THE BASICS

THE SIMPLEST AND MOST ENERGY-efficient way to make soap is the cold process method. The equipment needed is basic, and you can make soap pretty much anywhere you have access to water. Previously, I have used a method requiring access to a heat source, but in this book nearly all the soaps have been made by chopping cold fats into small pieces, adding any oils the recipe requires, and then pouring the freshly made lye solution directly onto the cold fats and oils. No heat source or thermometers are required. This is a huge time and energy saver that does not in any way diminish the quality of the finished soap.

If you prefer to melt your fats and oils over a heat source before adding your lye solution, wait until both the oils and the lye reach a similar temperature. Pouring boiling hot lye over slightly warmed oils can cause overheating, and in some instances, this can result in the mixture boiling up and overflowing (a bit like a volcano).

Traditional soapmakers in France use a hot process method, boiling the oils and lye together for several weeks before salting off the soap from the solution. There are more kitchen-friendly methods of hot processing your soap, and I have used these in two recipes on pages 60 and 63. One advantage of hot process soap is that the chemical reaction is complete once the soap is poured into molds, so it is immediately safe to use. While some curing time is still needed to allow the moisture in the soap to evaporate, this is not as lengthy as the curing time required in the cold process method.

Another soapmaking method is "melt and pour." While this method can produce colorful and artistic clear bar soaps, you have no control over your base ingredients since you simply buy a ready-made soap compound, melt it down, add color and fragrance, and pour into molds. As this book is about making products from scratch, I haven't included any "melt and pour" recipes.

EQUIPMENT

When I first set up my business, we somehow managed to make soap using plastic buckets, a handheld electric drill, and some large wooden boxes for molds. This is what makes soapmaking an industry that can be created in any outreach in the world. Here is what you will need for cold processed soapmaking:

Plastic buckets x 2 These will be used to mix up your lye solution and your soap. Domestic buckets are fine, but if you are buying new ones, buy rubbery rather than lightweight plastic because they will last longer.

Electric stick blender This speeds up the stirring process enormously. For large batches use an electric drill with a stainless steel or stripped metal paint stirrer attachment.

Plastic, polythene, or glass measuring jugs Buy half a dozen of these because you will find a million uses for them.

Stainless steel spoon Large with a long handle.

Molds and greaseproof paper Read more in the mold section on page 23.

Knife Large, nonserrated, and straight-edged for cutting fats and finished soaps.

Rubber gloves To protect your hands when handling sodium hydroxide and fresh soap.

Eye protection Perspex glasses or plastic goggles available from DIY shops.

Perspex glasses or plastic goggles available from DIY stores.

Sponges, paper towels, and dishcloths
For cleaning up during and afterward.

Kitchen scales
Digital scales weighing in small increments. You need specialist scales if you are making soap to sell.

Plastic beakers
For diluting colorants and weighing additives.

Droppers
Plastic or glass for adding colorants and essential oils.

Measuring spoons
For measuring out ingredients.

Kebab skewers, knitting needles, or a fork
For making swirls in your soap.

Plastic spatulas
For scraping the soap out of the bucket.

Sodium hydroxide (caustic soda/lye/NaOH)
Available in the plumbing section of DIY stores. This needs to be at least 95 percent pure.

Distilled or spring water
For mixing the lye.

Vinegar
If you splash yourself with lye, this will neutralize it.

Plastic wrap
For covering your freshly made soap.

Sugar thermometers x 2
Most of the recipes in this book use the very cold process method and do not require these, but, traditionally, you should measure the temperature of your oils and diluted lye and mix them together only when they reach the same temperature.

For hot processed soap and liquid soap making, you will also need a stainless steel stockpot and a larger pot to stand this in to create a bain-marie or double boiler. In addition, you will need a heat source hot enough to keep your soap at a rolling boil.

> NOTE: All pots and utensils that will come in contact with your soap must be made from stainless steel or plastic. Aluminium can react with the lye and flake off into your soap. You can use enamel but only if there are no chips on it. Avoid pots with a nonstick finish.

MOLDS

In theory you can use any wooden, plastic, rubber, or cardboard container as a mold for your cold or hot processed soap, as long as it will not leak. If it is inflexible, you must line the mold with baking parchment or plastic wrap so you can release your soap easily when it is set.

In practice, the speed of the saponification process and the quality of your finished soap can be influenced by the size of your mold—the larger the mold/soap batch, the greater the heat generated by the process. Block, tray, or loaf molds will give a more even soap than individual molds.

If you are making soap as a hobby, silicone baking molds are excellent. Since the soap does not stick to the surface, so you do not have to line your mold. You can also use cylindrical chip containers, flexible food storage containers, and even lengths of drainpipe to create curved shaped soaps (these should be greased with coconut oil or cooking spray before use).

If you are planning to sell your soap, there are a number of professional molds available that are sold with cutting devices to make it easy for you to ensure that each soap you make is the same size and weight. These molds are made either from wood or HDPE (high-density polyethylene)—a heat-resistant plastic that is easy to clean. You can also buy HDPE slab molds, which have plastic strips that divide the soap into equal-sized bars (as does the famous "Nizzy" mold, which Australian soapmakers swear by).

You should ensure that any mold you use is flexible. Clear, rigid plastic shapes are great for melt and pour soaps, but cold process soap does tend to get stuck to the bottom, and you may end up having to break the mold to get your soap out. Before resorting to this, try putting your soap-filled mold into the freezer for half an hour—this can help to release it.

> NOTE: The recipes in this book are all designed to fit a 3 lb (1½ kg) mold.

SAFE SOAP

Both bar and liquid soaps are made with sodium hydroxide (caustic soda, lye), which, without careful and sensible handling, can give you a nasty burn. If you take note of the following, all will be well:

· Store sodium hydroxide in an airtight and childproof container and label it "poison."

· Work in a well-ventilated space and mix lye outside if possible.

· Wear rubber gloves when handling lye or fresh soap.

· Always add lye to the water and not the other way around (think "LTW"—lye to water). If you pour water into your lye, it can bubble up and burn you.

· Wear a dust mask over your mouth and nose when adding the lye to the water (you only need to wear this for about two minutes while you stir the solution).

· Always wear goggles or protective glasses when handling lye and mixing your soap.

· When pouring dry lye into the water in the open air, hold the lye container inside and below the top edge of the bucket or a gust of wind can send it flying in all directions.

· Do not leave soap or your lye mix unattended or in the reach of children or pets.

· When making soap, keep a bottle of vinegar nearby. If you splash yourself with lye, rinse quickly with cold water and then apply the vinegar, which will help to neutralize the lye.

· Turn off your phone when mixing lye and weighing ingredients—you really must give it your full attention.

· Make sure your stick blender is immersed in your soap before you turn it on.

· When cleaning your soap bucket, first wipe it with paper and then put the paper in a sealed bag with your garbage. Wash all your utensils with regular dish soap after wiping the residue off with paper. (Fresh soap contains a lot of oil, and if you pour this down your sink on a cold day, it can solidify and block your drains.)

· Do not wear jewelry or keep a pin board or any small objects near your soapmaking area to avoid them falling into your soap.

· Cover your hair to avoid getting any in the soap.

SAPONIFICATION, SUPERFATTING, AND ESSENTIAL STUFF

SAPONIFICATION IS WHAT HAPPENS WHEN you mix oils or fats with lye. Every botanical or animal oil or fat has its own unique saponification (or SAP) value, and this value determines exactly how much sodium hydroxide is required to turn that particular oil into soap. Getting this right is the most important factor in all methods of soapmaking and is the difference between producing a harsh, potentially caustic bar of soap and a gentle moisturizing bar.

On the chart opposite, you will see a list of oils and two columns of numbers. The first column gives the SAP value for NaOH (sodium hydroxide) that should be used when making bar soaps. The second column gives the SAP value for KOH (potassium hydroxide) and should be used when making liquid soaps. The calculation is simple—multiply the quantity of oil or fat you want to use (in grams) by its SAP value to give you the quantity of lye needed to turn it into soap.

For example, to make soap using 20 oz (500 g) of coconut oil, multiply the NaOH SAP value for coconut oil (0.185) by 500 (0.185 x 500 = 92.5). This will tell you that you need 3¾ oz (92 g) of caustic soda to turn 20 oz (500 g) of coconut oil into soap (round any fractions up or down to the nearest whole number). For liquid soap, do exactly as above, but use the SAP value for KOH. If your recipe has several different oils or fats in it, calculate each one separately and then total the caustic soda up at the end. There are numerous online saponification programs that will do this for you (see Resources on page 143).

SUPERFATTING

Once you have calculated the amount of caustic soda needed, you can ensure that the soap will be gentle by superfatting it, or applying a "lye discount." Both of these terms mean exactly the same thing: either adding up to 5 percent extra oils or reducing the amount of lye by up to 5 percent. Deciding on what percentage is a little more complex because it is determined by the stability of your oils. If you are using an oil with a tendency to go rancid quickly, only superfat by up to 2 percent. If you want your soap to be extra oily or creamy, go up to 5 percent. This comes with experience, but for now, if you want to superfat a soap by say 3 percent (which is fine for most formulas) do your basic calculations to establish how much lye is needed and then deduct 3 percent from that amount.

When I first started soapmaking, I learned to add between 2 and 5 percent of a precious oil to use as a superfatter at the trace stage. Over the years, the logic of this has made less and less sense to me. Think about it: your soap takes up to six weeks to fully saponify—how do the soap fairies know which oil in the batch they should leave unsaponified? For many years now, I have been putting all of the oils together in the soap pot (including my superfatting oil), calculating the lye, and deducting the superfat percentage, and then adding the lye to the oils, which makes absolutely no difference in the quality of my finished soaps, so this is the method I now recommend using.

WATER CONTENT

The amount of water used in a formula is widely disputed among soapmakers, but I have found that using 33 percent of the total weight of the oils works best because it provides an even mix and gives me time to add my colors, herbs, and essential oils without panicking. Remember to always use distilled or spring water when making soap.

SAPONIFICATION CHART

OILS AND FATS	SAP NaOH	SAP KOH	OILS AND FATS	SAP NaOH	SAP KOH
Almond	0.136	0.191	Lecithin	0.086	0.121
Apricot Kernel	0.135	0.189	Macadamia	0.139	0.195
Avocado	0.133	0.187	Maize (corn)	0.136	0.191
Babassu	0.175	0.245	Mango butter	0.134	0.188
Baobab	0.142	0.245	Meadowfoam	0.121	0.170
Bay berry (Laurel)	0.141	0.180	Neem	0.139	0.195
Beeswax	0.069	0.097	Niger seed	0.136	0.191
Borage / Starflower	0.136	0.191	Nutmeg butter	0.180	0.252
Brazil nut	0.175	0.245	Olive	0.134	0.188
Camelia/teaseed	0.136	0.191	Palm	0.141	0.198
Candelilla	0.042	0.059	Palm kernel	0.164	0.230
Canola (Rapeseed)	0.124	0.174	Peach kernel	0.137	0.192
Carnauba wax	0.057	0.080	Peanut (Groundnut)	0.136	0.191
Castor	0.129	0.181	Pecan	0.135	0.189
Cocoa butter	0.137	0.192	Poppyseed	0.138	0.194
Coconut	0.185	0.260	Pumpkinseed	0.134	0.188
Cod-liver oil	0.130	0.182	Pura	0.136	0.191
Corn (Maize)	0.136	0.191	Rapeseed (Canola)	0.124	0.174
Cottonseed	0.134	0.188	Ricebran oil	0.128	0.180
Emu	0.137	0.192	Rosa mosqueta/rosehip	0.138	0.194
Evening primrose	0.136	0.191	Safflower	0.136	0.191
Fish oils	0.133	0.187	Sesame seed	0.133	0.187
Flax	0.136	0.191	Shortening	0.138	0.194
Goose fat	0.137	0.192	Shea butter (Karite)	0.128	0.180
Grapeseed	0.127	0.178	Soybean (Soya)	0.135	0.189
Groundnut (peanut)	0.136	0.191	Stearic acid	0.143	0.200
Hazelnut	0.136	0.191	Sunflower	0.134	0.188
Hempseed	0.135	0.189	Talllow, bear	0.139	0.195
Mafura Butter	0.200	0.194	Tallow, beef	0.140	0.196
Jojoba	0.068	0.095	Tallow, goat,sheep,deer	0.140	0.196
Marula Oil	0.135	0.189	Walnut	0.135	0.189
Lanolin	0.074	0.104	Watermelon (Kalahari)	0.133	0.187
Lard	0.138	0.194	Wheat germ	0.131	0.180

CURING

When making cold processed soaps, the saponification process can take up to six weeks to complete. In addition to this, to create a hard, long-lasting soap bar, the water used in the process needs time to evaporate, so soaps need to be cured. To do this, lay your cut bars on a tray with spaces in between so the air can circulate. Place them somewhere dry (an airing cupboard is ideal) and leave them for six weeks, turning them from time to time. If you use your soap before the curing process is complete, it could irritate or sting your skin. How do I know when it's ready? The traditional method used by soapmakers is "the tongue test": wet your finger and rub it on the surface of your soap and then put that finger to your tongue. If it "bites" or "tingles," your soap needs longer to cure. If it just tastes like soap, it's ready. Once you have made a few batches of soap, you will know from the hardness of the bar when the soap is ready.

Soap made using the hot process method does not need time to cure, but it does need time for the water to evaporate. Likewise, liquid soaps are ready for use once the paste is diluted. The curing process can be reduced by half if you put your soaps in an enclosed space and keep a dehumidifier running 24/7. You do not need to cure your liquid soaps.

CUTTING YOUR SOAP

Soap can be cut into blocks using any straight-edged blade or cheese wire, but if you are selling your soap, it is important, for legal reasons, to ensure that the bars are all the same weight. There are simple devices available to cut your soap, or you can make something similar to a miter block with a straight slit in it that enables you to slide your soap block forward to the same position each time before cutting a chunk off. When the soap should be cut depends to some degree on whether you have used soft or hard oils in your formula, but it is normally between twenty-four and forty-eight hours after it comes out of the mold, by which time it should look and feel like

hard cheese. Be careful, though; if you leave it too long, the soap can become too hard to cut without crumbling.

INSULATING, GELLING, AND TROUBLESHOOTING

"Gelling" is a phase that an oil and lye combination sometimes goes through during the saponification process. A dark shadow will appear at the center of your soap, and over a period of hours, this shadow will spread out to the edges of your mold and then (usually over the next twenty-four hours) disappear. I say "sometimes" because the question of whether a soap will gel depends on the recipe formula, the water content, the size and shape of the mold, and the weather or ambient temperature you are working in (the warmer the environment and the larger the batch, the more likely your soap is to gel). Basically, what happens during this phase is that the saponification process speeds up.

There is much debate as to whether a gelled soap is superior to an ungelled soap (i.e., one that has saponified more slowly). Many soapmakers insulate their molds by wrapping blankets round them in an attempt to achieve gel—others put their freshly poured soaps in the fridge hoping to avoid it. On the plus side, gelling usually produces a hard bar of soap that cures relatively quickly, but on the minus side, quite often the shadow on the surface of the soap results in permanent patches of uneven color.

The most important thing to note is that if you attempt to cut your soap while it is in the gel phase, you will find yourself faced with a glutinous mess. So if a shadow appears on the surface of your soap, make sure it has faded away before you attempt any cutting.

Natural soapmaking is alchemy, and as such, we have to take into account some of the vagaries of nature. As an inexperienced soapmaker working with the dreaded caustic soda, you may look at your finished soaps and be thrown into long, nail-biting sessions while you decide whether they are safe to use. Here are some of the most common soap ailments and their most likely causes, but be aware that most mistakes only happen if you have failed to calculate your basic formula correctly.

Crumbly soap

Caused by too little water or excess lye. Grate this up and use it as a laundry soap.

Curdling

Nearly always happens with goat's milk soap and with liquid soaps, especially when using two different lyes. Just keep stirring, it will be all right.

Mottled soap

Often occurs in soaps colored with spices that haven't been stirred well enough, but fine to use.

No Trace

Soaps made without lauric (palm/coconut) fats can take much longer to trace. If you are sure your basic formula has the correct amount of lye, keep zapping with your hand blender.

Powdering

White powder on the top of your soap is actually soda ash—completely harmless but not very pretty. Scrape or wipe it off, and next time be sure you cover the surface of your soap mold with plastic wrap to prevent this.

Seizing or separation

You add your fragrance and suddenly the soap goes into shock and becomes a lump (or several lumps) of fat floating in a sea of oil. This is most commonly caused by compounds in your fragrance or essential oils, and you can stir the oils and fat together and scoop the lot into your mold. If the soap doesn't absorb the oil over time, throw it out because it is probably lye heavy.

Pockets of moisture

Bubbles filled with brown fluid on the surface of your soap. If you have honey in your recipe, this is probably the culprit. If you don't, these are possibly pockets of lye and you should discard the soap.

Soft soap

Soaps made with 100 percent liquid oils can take several weeks to harden, but if this is a 100 percent olive oil soap, it's worth the wait.

If this is a regular batch containing coconut oil, you have weighed your lye incorrectly. The soap won't hurt you, but it will go gloopy when wet.

Dreaded orange spot

A common soapmaking disease caused by oxidization of your oils. The surface of your soap will be covered with small, orangey-brown spots or bruises. While the soap is safe to use, throw it out because this condition will spread to neighboring soaps.

BASIC SOAP METHOD

1. Chop up the fats, and then weigh the fats and base oils one by one and put them together in a plastic bucket.

2. Weigh the water in the second bucket.

3. Wearing goggles and gloves, weigh the sodium hydroxide in a jug.

4. Line or grease molds.

5. Weigh additives, essential oils, and colorants.

6. If using several essential oils, blend these together and add any herbs or dried additives you may be using.

7. Wearing your gloves, goggles, and mask, go to a well-ventilated area and pour the sodium hydroxide into the water. Stir with a long-handled, stainless steel spoon until dissolved. The chemical reaction will cause the solution to heat up, and you will see steam and fumes rising from the bucket.

1

8

10

9

12

8. Pour the hot lye directly over the cold oils and stir until the oils and fats have dissolved.

9. Zap the mixture with a stick blender until it traces (zap in short bursts so the motor doesn't burn out).

10. Add herbs and/or colorants and stir in thoroughly with a spoon.

> NOTE: "Trace" is the point when soap dribbled from a spoon back onto the surface of your batch will form a line. In "light" trace this is just distinguishable, "medium" trace it is quite clear, and "heavy" trace it is prominent.

11. Add your essential oils if they are not already combined with previous ingredients. Stir well.

12. Pour your soap into your prepared molds.

13. Place plastic wrap directly on the surface of your soap and leave for 48 hours.

14. Turn the soap out of the mold and cut into bars using a sharp straight-edged knife, a cutting blade, or cheese wire.

15. Place your soaps on a tray (leaving spaces in between) and put them in a dry place for six weeks.

16. Wash and enjoy!

PRESENTATION

With each of the recipes in this book, I have given you suggestions for packaging. Soap is a truly versatile product that can be squished, molded, and cut into many forms and fantasies—the only limitation you have is your imagination. When you are trying to come up with new ideas, the first thing you should do is stop thinking of a bar of soap as just square or oblong.

For starters, soap on a rope can be many things. All you need is an electric drill to make a clean hole through any bar and thread a cord through it. Knot it at the bottom to secure.

It is worth remembering that all but the most discerning customers buy first because of presentation and second because of the quality of the product itself. Ironically, the simpler your packaging (think Chanel white boxes), the classier the positioning and the higher the price you can demand. When designing your soaps, think of upmarket bathrooms and what will look good displayed in them. Almost 60 percent of soaps sold in the gift market are never used—they just sit for years in the bathroom looking pretty.

PURPOSE

While we think of soap primarily as a product to wash ourselves with, it can be packaged and sold in numerous other ways. Heavily scented natural soap will fragrance a room and will certainly fragrance clothes. You can capitalize on this by cutting your oddments into interesting shapes and packaging them as drawer fresheners, or you could tie a ribbon through them so they can be hung on clothes hangers. If you use insect-repelling essential oils in your soaps (such as citronella, rosemary, or lemongrass), the drawer fresheners will also repel insects and moths. Selling soap as laundry detergent bypasses all the lengthy cosmetic legislation and is a great way to package up soap offcuts. Simply grate the soap and package it in a bag—this could be a plain paper sack with a scoop or muslin bag that can just be thrown in the washing machine.

Soap can also be molded or cut into cubes to form beads. It can be carved with a craft knife (the Japanese carve wonderful flowers from it), and it can even be cut into small tiles which can then be used to create mosaics. Wet soap is a gluey substance; if you score the back of the tiles and wet them, you can press them onto paper and they will stick firmly to it.

DECORATION

Soap can be decorated in so many ways. The important thing is to forget that it is a bar of soap. Items such as shells, seaweed, and even money can be embedded in the top—wet the surface just after cutting and push them in. If you want to follow a popular trend and stud your soap with pink rosebuds, it's a good idea to dip the rosebuds in candle wax first. The problem with untreated botanicals is that when the soap gets wet, the botanicals turn brown and soggy, and even a small amount of wetness may cause the flowers/plants to go moldy. Pure food-grade gold leaf can be pressed onto the surface of freshly cut soap to great effect, and powdered mica (which does not hold up well when used as a colorant) will add a metallic or pearlized sheen when sprinkled on the top of a newly cut bar.

Paper decals can be embedded in the surface of a soap, and you can even buy soap paints and stencil names or initials onto your bars. You can also achieve this by pouring fresh soap onto a stencil placed on a ready cut bar. Cake decorations (hundreds and thousands of silver balls) can be used to decorate your soaps, and silk flowers, millinery decorations, and many items, perhaps originally bought for scrapbooking, can be applied either to the soap itself or to your packaging.

If you are looking for an idea, just go to your nearest main street and think soap!

PURE AND GENTLE

AN INCREASING NUMBER OF PEOPLE SEEM to be suffering from allergies and skin problems. While some people are quite plainly allergic to pure soap products, they are vastly outnumbered by those who simply cannot tolerate synthetic detergent bars. The cold process soaps in this book contain pure glycerine, which helps protect and moisturize the skin, but since fragrances (natural or otherwise) and colorants contribute greatly to skin problems, you need to be extremely careful about what you add to your soaps (and creams) if you are trying to create a product for allergy-prone skin.

While there are a limited number of essential oils that can be used in small quantities in your pure and gentle soaps, it is best to keep soaps for allergy-prone skin as simple as possible. This section contains the best recipes I can come up with for those suffering from eczema and related problems. The soaps are packed with soothing ingredients and superfatted by 5 percent to ensure no lye can survive.

Purity

This super mild soap is suitable for use on babies and on your face. It has no color or fragrance and has shea butter for maximum moisturizing qualities. It is superfatted by 5 percent and so includes vitamin E oil to help preserve the free oils from rancidity.

INGREDIENTS

9½ oz (240 g) coconut oil

8 oz (200 g) shea butter

16 oz (400 g) sweet almond oil

4 oz (100 g) borage oil

2 oz (50 g) apricot kernel oil

13 oz (327 g) spring or distilled water

5½ oz (138 g) sodium hydroxide

2 tsp (10 ml) vitamin E oil

Superfatted at 5%

1 Weigh the coconut oil and shea butter, cut into small chunks, and place in a plastic bucket. Weigh and add the almond, borage, and apricot oils.

2 Put on your rubber gloves and goggles. Weigh the water in a plastic bucket. Weigh the sodium hydroxide in a jug.

3 In a well-ventilated area and while wearing gloves, mask, and goggles, pour the sodium hydroxide into the water, stir with a long-handled, stainless steel spoon until dissolved, and then set aside. Remove mask.

4 Pour the caustic solution into the oils. Stir manually until the fats and oils have dissolved and then bring the soap to a light trace using a stick blender.

5 Add the vitamin E oil. Mix in thoroughly with either a spoon or a blender and pour into a prepared 3 lb (1½ kg) mold.

6 Leave the soap to set for approximately 48 hours or until it has reached a "hard cheese" consistency. Turn out of the mold and cut into bars. Cure the bars for four to six weeks before use.

VARIATION

For mature skin, deduct 1 tsp almond oil from your formula and replace this with 1 tsp evening primrose oil added at trace.

PACKAGING IDEA

For a perfect baby gift, package in a white box with a clear lid together with the Baby's Bottom Cream on page 117 and tie with a blue or pink ribbon.

Calendula and Goat's Milk

Goat's milk is a deceptive ingredient. It looks and smells horrid when you mix it with the lye, but it makes a beautiful creamy soap that is mild. I've added a dash of chamomile and lavender essential oils here since both are safe to use on sensitive skin.

1 Weigh the coconut oil and cocoa butter, cut into small chunks, and put in a plastic bucket. Weigh and add the olive and avocado oils. Measure out the essential oils and blend them together.

2 Put on your rubber gloves and goggles. Weigh the water in a plastic bucket and the goat's milk in a jug and mix together in the bucket. Weigh the sodium hydroxide in a jug.

3 In a well-ventilated area and while wearing gloves, mask, and goggles, pour the sodium hydroxide into the water and milk mixture and stir with a long-handled, stainless steel spoon until dissolved. The goat's milk may curdle and turn yellow. It will also throw off fumes that smell of ammonia, but continue stirring and keep your mask on. When the mixture is as even as you can get it, set aside. Remove mask.

4 Pour the caustic solution into the oils. Stir manually until the fats and oils have dissolved and then bring the soap to a light trace using a stick blender. Add the essential oil blend and stir thoroughly.

5 Sprinkle in the calendula petals—the amount you use is not critical, just enough to make the soap look pretty. Mix in thoroughly with a spoon and pour into a prepared 3 lb (1½ kg) mold.

6 Leave the soap to set for approximately 48 hours or until it has reached a "hard cheese" consistency. Turn out of the mold and cut into bars. Cure the bars for four to six weeks before use.

INGREDIENTS

12 oz (300 g) coconut oil

2 oz (50 g) cocoa butter

24 oz (600 g) olive oil

2 oz (50 g) avocado oil

6 oz (165 g) spring or distilled water

6 oz (165 g) full-fat goat's milk

5½ oz (142 g) sodium hydroxide

A handful of calendula petals

ESSENTIAL OILS

½ tsp (2.5 ml) Roman chamomile

½ tsp (2.5 ml) lavender

Superfatted at 5%

VARIATION

To create a soap with antiseptic properties, replace the chamomile and lavender essential oils with a blend of tea tree and lemon essential oils.

PACKAGING IDEA

Cut the base of this soap with a wavy-edged herb cutter to help the soap to drain and stop it going gooey, then tie it to a wooden soap dish with raffia.

Green Clay and Oatmeal

This gentle layered soap is great for exfoliating, and the green clay will draw impurities from the skin. I've included a small percentage of castor oil here since it is a powerful antioxidant that conditions the skin, but be aware that soaps containing castor oil trace very quickly.

INGREDIENTS

16 oz (400 g) coconut oil
16 oz (400 g) olive oil
4 oz (100 g) castor oil
2 oz (50 g) rosehip seed oil
2 oz (50 g) sunflower oil
½ oz (14 g) green clay
½ oz (14 g) oatmeal
13 oz (330 g) spring or distilled water
5¾ oz (146 g) sodium hydroxide

ESSENTIAL OILS

½ tsp (2.5 ml) rose geranium
½ tsp (2.5 ml) lemon

Superfatted at 5%

1 Weigh the coconut oil, cut into small chunks, and put in a plastic bucket. Weigh and add the olive, castor, rosehip seed, and sunflower oils. Weigh the green clay and oatmeal.

2 Put on your rubber gloves and goggles. Weigh the water in a plastic bucket. Weigh the sodium hydroxide in a jug. Add just enough water from the weighed quantity to the green clay to make it fluid.

3 In a well-ventilated area and while wearing gloves, mask, and goggles, pour the sodium hydroxide into the water, stir with a long-handled, stainless steel spoon until dissolved, and then set aside. Remove mask.

4 Blend the essential oils together and pour half onto the oatmeal and the other half onto the green clay. Pour the caustic solution into the oils. Stir manually until the fats and oils have dissolved and then bring the soap to a light trace using a stick blender.

5 Divide the soap equally between two jugs. Add the green clay to one jug and the oatmeal to the other and stir thoroughly. Zap the soap containing the green clay with a blender until it is at heavy trace and pour it into the molds (if using individual molds, pour to the halfway mark). Carefully pour the oatmeal soap over the top.

6 Leave the soap to set for approximately 48 hours or until it has reached a "hard cheese" consistency. Turn out of the mold and cut into bars. Cure the bars for four to six weeks before use.

VARIATION

For a really scrubby foot soap, replace the oatmeal with coarse pumice powder, increasing the quantity so that the bottom layer is almost solid pumice.

PACKAGING IDEA

Plait several strands of raffia together to form two ropes. Stack three soaps together and tie the two raffia ropes round the soaps, knotting at the top.

FRUIT SOAPS

FRUIT-FLAVORED SOAPS SELL WELL, WHICH may be a problem for those of us who strive to keep our soaps entirely "natural" since, with the exception of most of the citrus varieties, few fruits are available as pure essential oils. If you want true fruit smells, you have to resort to using artificial fragrance oils (some good ones designed specifically for cold process soap are available).

You can, of course, gain the unquestionable benefits of fruit (vitamins, omega acids, etc.) by including fruit oils and extracts in your soap, but these will not give your soap fragrance or color. There are also powdered fruit extracts on the market, and you can use dried, powdered citrus peel, which is not only exfoliating but also bursts into numerous tiny yellow starlets that will decorate the surface of your soap.

Another option is to add liquidized fresh fruit to your soap, although lumps of fruit on the surface will attract bacteria, so ensure that the mixture is completely lump free. When using fresh fruits, you should always include an antioxidant such as vitamin E oil. While you can, add liquidized fruit at trace. To ensure a long shelf life, add it to the lye and deduct from the weight of your water.

Fruit Swirls

Swirled soaps are easier to achieve in a tray mold with a large surface area. Swirling in small individual molds can be difficult because the soap may go hard before you are halfway through the batch.

INGREDIENTS

16 oz (400 g) coconut oil

12 oz (300 g) palm oil

12 oz (300 g) sunflower oil

13 oz (330 g) spring or distilled water

6 oz (148 g) sodium hydroxide

1 Tbsp (15 ml) fruit fragrance oil

1 tsp (5 ml) diluted oxide

Superfatted at 3%

1 Weigh the coconut and palm oil, chop them into small chunks, and put them in a plastic bucket. Weigh and add the sunflower oil.

2 Put on your rubber gloves and goggles. Weigh the water in a plastic bucket. Weigh the sodium hydroxide in a jug.

3 In a well-ventilated area and while wearing gloves, mask, and goggles, pour the sodium hydroxide into the water, stir with a long-handled, stainless steel spoon until dissolved, and then set aside. Remove mask.

4 Pour the caustic solution into the oils. Stir manually until the fats and oils have dissolved and then bring the soap to a light trace using a stick blender.

5 Add the fragrance oils. Mix in thoroughly with either a spoon or blender. If you plan to use a fruit extract add it now.

6 Divide the batch into two jugs, ⅔ in one jug, ⅓ in the other. Pour the colorant into the jug containing the smaller batch of soap and mix thoroughly.

7 When the soap is at a heavy trace, pour a layer of the uncolored soap (approximately ⅓ of the jug) into the mold. Dribble lines of soap from the colored batch over the surface. Trail your pointy object through the colored soap to form the swirly pattern of your choice.

8 Repeat step 7 twice, building up layers of soap and swirling as you go until all your soap is in the mold. Tap the sides of the mold to level out the surface.

9 Leave the soap to set for approximately 48 hours or until it has reached a "hard cheese" consistency. Turn out of the mold and cut into bars. Cure the bars for four to six weeks before use.

VARIATION

If you prefer to keep your soaps completely natural, use annatto as a colorant instead of using the oxide. You can also replace the fragrance oil with 1 Tbsp (15 ml) lemon, grapefruit, sweet orange, or mandarin essential oils to give you a pure, natural scent.

PACKAGING IDEA

Fruit soaps are great packed in little wooden crates filled with wood wool. You will find the crates in many craft outlets, or just do a search on the Internet for "soap gift crates."

Fig and Honey

Figs are one of the latest buzz ingredients to be included in soap and cosmetic ranges. They are packed with vitamin A and act as a humectant when combined with honey, attracting moisture to the skin. The seeds are also great natural exfoliants.

INGREDIENTS

20 oz (500 g) coconut oil

6 oz (150 g) shea butter

14 oz (350 g) olive oil

13 oz (330 g) spring or distilled water

6 oz (154 g) sodium hydroxide

2 medium fresh figs

1 Tbsp (15 ml) honey

1 tsp (5 ml) vitamin E oil

ESSENTIAL OILS

2 tsp (10 ml) ylang-ylang

½ tsp (2.5 ml) cinnamon

½ tsp (2.5 ml) ginger

Superfatted at 3%

1 Weigh the coconut oil and shea butter, cut into small chunks, and place in a plastic bucket. Weigh and add the olive oil.

2 Put on your rubber gloves and goggles. Weigh the water in a plastic bucket. Weigh the sodium hydroxide in a jug.

3 In a well-ventilated area and while wearing gloves, mask, and goggles, pour the sodium hydroxide into the water, stir with a long-handled, stainless steel spoon until dissolved, and then set aside. Remove mask.

4 Liquidize the figs using a stick blender, leaving no lumps. Pour the liquidized figs and honey into the caustic solution and stir until dissolved. Blend the essential oils together.

5 Pour the caustic solution into the oils. Stir manually until the fats and oils have dissolved and then bring the soap to a light trace using a stick blender. Add the vitamin E and essential oils. Combine thoroughly with either a spoon or your blender and pour into a prepared 3 lb (1½ kg) mold.

6 Leave the soap to set for approximately 48 hours or until it has reached a "hard cheese" consistency. Turn out of the mold and cut into bars. Cure the bars for four to six weeks before use.

VARIATION

If you are using green figs, try adding a tablespoon of diluted green oxide at trace, and if you can't find fresh figs, add 1 percent fig extract at trace.

PACKAGING IDEA

If you are lucky enough to have access to fresh fig or vine leaves, package your soap in a parcel made from a fresh leaf and tie it up with raffia.

Apricot Skin Softener

Apricot kernel oil is rich in fatty acids and vitamin E (a natural antioxidant), and is easily absorbed by the skin. This is one of the key oils to use in soaps, creams, and lotions for those with sensitive skin because it is deeply nourishing.

1 Weigh the coconut oil and shea butter, cut into small chunks, and put in a plastic bucket. Weigh and add the olive, almond, and apricot oils.

2 Put on your rubber gloves and goggles. Weigh the water in a plastic bucket. Weigh the sodium hydroxide in a jug.

3 In a well-ventilated area and while wearing gloves, mask, and goggles, pour the sodium hydroxide into the water, stir with a long-handled, stainless steel spoon until dissolved, and then set aside. Remove mask.

4 Blend the essential oils together and pour them onto the turmeric. Add enough water to make the consistency of a pourable paste.

5 Pour the caustic solution into the oils. Stir manually until the fats and oils have dissolved and then bring the soap to a light trace using a stick blender.

6 Pour two tablespoons of soap into the turmeric and mix until smooth. Return this mixture back into the main batch and zap with your blender until you have reached trace. Pour this soap into your mold. (Alternatively, use cupcake papers, as in the photograph opposite, and pour the soap at medium trace.)

7 Leave the soap to set for approximately 48 hours or until it has reached a "hard cheese" consistency. Turn out of the mold and cut into bars. Cure the bars for four to six weeks before use.

INGREDIENTS

16 oz (400 g) coconut oil

6 oz (150 g) shea butter

8 oz (200 g) olive oil

8 oz (200 g) almond oil

2 oz (50 g) apricot kernel oil

13 oz (330 g) spring or distilled water

5½ oz (150 g) sodium hydroxide

2 Tbsp turmeric

ESSENTIAL OILS

2 tsp (10 ml) geranium

½ tsp (2.5 ml) sweet orange

½ tsp (2.5 ml) mandarin

Superfatted at 3%

VARIATION

For really sensitive skin, try a blend of rose, lemon, and Roman chamomile essential oils using 2 tsp (10 mls) total instead of 3 tsp (15 mls).

PACKAGING IDEA

To emphasize the gentleness of this soap, wrap it in handmade silk paper and tie it with silk ribbon.

SCRUB BARS

THE LIST OF INGREDIENTS YOU CAN PLAY with when making soap is virtually endless. In this section, the soaps are designed around texture and purpose. You can choose mild exfoliation by adding botanicals such as dried herbs or rose petals, or for heavy scrubbing, try adding coarse pumice, oatmeal, or even sand.

The amount of exfoliating material you put in your soap is not critical—add as much or as little as you like—you can also replace one exfoliant with another. Why not experiment with the many grains you can find at the supermarket? Once you begin to think like a soapmaker, no food stuff, container, or plant is safe!

Poppy seeds are not only decorative but are useful as an exfoliant, and you could also line your mold with a layer of tapioca or round beans to give you a massage bar. Try fruit seeds and ground kernels—strawberry and raspberry seeds work well and look pretty, and ground olive stones are a vitamin-rich additive.

Rose and Oatmeal

This bar is packed with the goodness of dried rose petals and bran. Rose petals can be bought ready-dried, but it is a good idea to put them through a coffee grinder before use because whole petals can go slimy in your soap.

INGREDIENTS

20 oz (500 g) coconut oil

7½ oz (190 g) shea butter

12 oz (300 g) rapeseed (canola) oil

½ oz (10 g) wheat germ oil

13 oz (330 g) spring or distilled water

1 Tbsp (15 g) powdered madder

6½ oz (151 g) sodium hydroxide

½ oz (15 g) ground oatmeal

½ oz (8 g) ground rose petals

2 tsp (10 ml) vitamin E oil

Dried rosebuds, for decoration

ESSENTIAL OILS

1 tsp (5 ml) geranium

½ tsp (2.5 ml) patchouli

½ tsp (2.5 ml) sweet orange

Superfatted at 3%

1 Weigh the coconut and shea butter, cut into small chunks, and put in a plastic bucket. Weigh and add the rapeseed and wheat germ oils.

2 Put on your rubber gloves and goggles. Weigh the water in a plastic bucket. Take 3½ fl oz (100 ml) water from the bucket, heat it slightly, and pour it over the powdered madder. Weigh the sodium hydroxide in a jug.

3 Weigh out your oatmeal and your rose petals. Blend your essential oils and put them in with the oatmeal.

4 In a well-ventilated area and while wearing gloves, mask, and goggles, pour the sodium hydroxide into the water, stir with a long-handled, stainless steel spoon until dissolved, and then set aside. Remove mask.

5 Pour the caustic solution into the oils. Stir manually until the fats and oils have dissolved and then bring the soap to a light trace using a stick blender.

6 Add the vitamin E oil. Add the madder paste a spoonful at a time until you achieve the desired color. Add the rose petals and oatmeal and mix in thoroughly with either a spoon or a blender and pour into prepared 3 lb (1½ kg) or individual molds.

7 Leave the soap to set for approximately 48 hours or until it has reached a "hard cheese" consistency. Turn out of the mold and cut into bars. Cure the bars for four to six weeks before use.

VARIATION

Directly after your bars are cut, push whole rosebuds into the surface of the soap. It is a good idea to dip these in melted candle wax first.

PACKAGING IDEA

Sprinkle loose rose petals over the surface of the soap and wrap it in clear cellophane.

Poppy Seed

I love using poppy seeds both as an exfoliant and for decoration. The heart mold
I have used for this soap is a small silicone cake mold that I filled with
a basic poppy seed soap and then put some seeds on the top.

1 Weigh the coconut and palm oils and shea butter, chop into small
chunks, and put in a plastic bucket. Weigh and add the olive oil.

2 Put on your rubber gloves and goggles. Weigh the water in a plastic
bucket. Weigh the sodium hydroxide in a jug.

3 Weigh the poppy seeds. Blend the essential oils and put them in with
the poppy seeds. (If you are making a layered soap, keep half the
essential oil back.)

4 In a well-ventilated area and while wearing gloves, mask, and goggles,
pour the sodium hydroxide into the water, stir with a long-handled,
stainless steel spoon until dissolved, and then set aside. Remove mask.

5 Pour the caustic solution into the oils. Stir manually until the fats
and oils have dissolved and then bring the soap to a light trace using
a stick blender. If you are making a layered soap, divide the batch
into two jugs.

6 Add the colorant, drop by drop, into one of these jugs until the desired
color is achieved. Now add the remaining essential oil and zap with a
blender until the soap reaches a heavy trace. Pour this soap into the
mold.

7 Add the poppy seeds to the second jug, mix in thoroughly with either
a spoon or a blender, and pour into prepared molds, or over the first
layer of blue soap.

8 Leave the soap to set for approximately 48 hours or until it has reached
a "hard cheese" consistency. Turn out of the mold and cut into bars.
Cure the bars for four to six weeks before use.

INGREDIENTS

16 oz (400 g) coconut oil

12 oz (300 g) palm oil

4 oz (100 g) shea butter

8 oz (200 g) olive oil

13 oz (330 g) spring or
distilled water

6 oz (151 g) sodium hydroxide

⅓ oz (8 g) poppy seeds

½ tsp (2.5 ml) diluted
ultramarine blue oxide (optional)

ESSENTIAL OILS

1 tsp (5 ml) ylang-ylang

½ tsp (2.5 ml) howood

¼ tsp (1 ml) sweet orange

¼ tsp (1 ml) palmarosa

Superfatted at 3%

VARIATION

Replace the blue oxide with 3 Tbsp of cocoa powder mixed with
a little water into a smooth paste. Use sweet orange,
cinnamon, and ginger essential oils.

PACKAGING IDEA

If you are cutting this soap into regular bars, place a slice of dried orange
on top of the soap and secure with a raffia tie.

Gardeners' Soap

If you would prefer to use vegetable oils rather than lard to make a gardeners' soap, use any of the bar soap recipes in this book and substitute any exfoliant additives with pumice, using the essential oil blends suggested below.

INGREDIENTS

24 oz (600 g) lard

16 oz (390 g) rapeseed (canola) oil

½ oz (10 g) wheat germ oil

13 oz (330 g) spring or distilled water

4¾ oz (128 g) sodium hydroxide

coarse pumice

½ oz (10 g) vitamin E oil

ESSENTIAL OILS

1 tsp (5 ml) rosemary

½ tsp (2.5 ml) tea tree

½ tsp (2.5 ml) lavender

Superfatted at 3%

1 Weigh the lard, chop into small chunks, and put in a plastic bucket. Weigh and add the rapeseed and wheat germ oils.

2 Put on your rubber gloves and goggles. Weigh the water in a plastic bucket. Weigh the sodium hydroxide in a jug. Weigh out the pumice. Blend the essential oils and add to the pumice.

3 In a well-ventilated area and while wearing gloves, mask, and goggles, pour the sodium hydroxide into the water, stir with a long-handled, stainless steel spoon until dissolved, and then set aside. Remove mask.

4 Pour the caustic solution into the oils. Stir manually until the fats and oils have dissolved, and then bring the soap to a light trace using a stick blender.

5 Add the vitamin E oil. Add the pumice and mix in thoroughly with either a spoon or a blender—you want a thick mixture for a scrubby bar so add more pumice if the mix looks too thin. Pour into prepared 3 lb (1½ kg) molds.

6 Leave the soap to set for approximately 48 hours or until it has reached a "hard cheese" consistency. Turn out of the mold and cut into bars. Cure the bars for four to six weeks before use.

VARIATION

Try substitututing the pumice with cut loofah and the essential oils for a blend of lavender and lime.

PACKAGING IDEA

Cover a cardboard box with cut-up seed packets and put a soap and a Gardeners' Balm (see page 128) inside to make a great gardeners' gift.

PROVENÇAL

WHEN IT COMES TO SOAP INGREDIENTS, the French are truly blessed. A visit to Provence, with its acres of stunning lavender fields, will show you just how it got its reputation for wonderful fragrances. I live in the Dordogne, a southwest region of France, and introducing handmade soaps to the French is a little like trying to sell refrigerators to Eskimos. Soapmaking is so much part of the culture that it is hard to convince the French that there are alternative methods to the seven days of boiling and salting used to create their most famous soap, Savon de Marseille, originally made from pure olive oil.

The original Savon de Marseille soaps were unperfumed and uncolored. Here, I give you my version of the original castile (olive oil) soap, which I proudly name Savon de la Ste Sabine (see page 60). I have used the hot process method to make this soap, as well as Laurus Nobilis (see page 63), my version of Savon d'Alep (a traditional soap made with bay/laurel berry oil).

Lavender Provençal

Lavender outsells all varieties of soap by eight to one. Whole ranges of soaps and toiletries can be created using just this one plant, and here I divide one batch into three to give you a variety of ways of presenting this perennial favorite.

INGREDIENTS

18 oz (450 g) coconut oil

20 oz (500 g) olive oil

2 oz (50 g) beeswax (flakes or granules)

13 oz (330 g) spring or distilled water

6 oz (148 g) sodium hydroxide

½ oz (14 g) dried lavender buds

2 tsp (10 ml) vitamin E oil

4 full lavender heads

2 tsp (10 ml) each blue and mauve oxide, diluted

ESSENTIAL OILS

1 Tbsp (15 ml) French lavender

Superfatted at 4%

1. Weigh the coconut, olive oil, and beeswax and put in a stainless steel saucepan on a low heat to melt. Remove from heat as soon as melting point is reached.

2. Put on your rubber gloves and goggles. Weigh the water in a plastic bucket. Weigh the sodium hydroxide in a jug. Weigh the lavender buds and measure out the essential oils.

3. In a well-ventilated area and while wearing gloves, mask, and goggles, pour the sodium hydroxide into the water, stir with a long-handled, stainless steel spoon until dissolved, and then set aside. Remove mask.

4. Check the temperature of both the caustic solution and the oils, and when both reach approximately the same temperature, pour the caustic solution into the oils. Stir manually for three minutes and then bring the soap to a light trace using a stick blender.

5. Add the vitamin E oil and the essential oil and stir to medium trace. Divide the batch equally between three jugs. Add the blue colorant drop by drop into one jug until the desired color is achieved. Pour the soap into a mold.

6. Add the mauve colorant and the lavender buds to the second jug. Stir thoroughly and pour into a second mold. Leave the third batch plain—pour into the remaining mold then place the whole lavender heads on the top for decoration.

7. Leave the soap to set for approximately 48 hours or until the soap has reached a "hard cheese" consistency. Turn out of the molds and cut into bars. Cure the bars for four to six weeks before use.

VARIATIONS

To make a layered soap, add the colorants to each jug of soap in step 5. Bring the first jug to heavy trace with a blender and pour into your mold. Zap the second jug and pour this soap on the top. Pour the third layer on top without zapping.

PACKAGING IDEA

Place sprigs of lavender on top of the soap and wrap in cellophane. Alternatively package one of each soap in a white box with a clear acetate lid and tie with lavender-colored ribbon and a tag. Soap shavings tied in muslin fabric can be sold as drawer fresheners.

Argile Clay

Originating in France, argile (or montmorillionite) clay is a kaolin powder that draws toxins out of the skin. Available in green, red, rose, yellow, and white, it also acts as a totally natural colorant and gives your soap a silky texture.

1 Weigh the coconut oil and shea butter, cut into small chunks, and put in a plastic bucket. Weigh and add the rapeseed and olive oils.

2 Put on your rubber gloves and goggles. Weigh the water in a plastic bucket. Weigh the sodium hydroxide in a jug. Weigh the three clays. Measure out and blend together the essential oils, divide the blend in three, and add to the clays in equal amounts. Add enough water to each clay to make a pourable paste.

3 In a well-ventilated area and while wearing gloves, mask, and goggles, pour the sodium hydroxide into the water, stir with a long-handled, stainless steel spoon until dissolved, and then set aside. Remove mask.

4 Pour the caustic solution into the oils. Stir manually until the fats and oils have dissolved and then bring the soap to a light trace using a stick blender.

5 Divide the batch equally between three jugs. Add one clay color to each jug. Working quickly, zap the first jug with the blender, taking care that the clay is distributed evenly. Pour into a prepared mold. Repeat for the other two jugs.

6 Leave the soaps to set for approximately 48 hours or until the soap has reached a "hard cheese" consistency. Turn out of the molds and cut into bars. Cure the bars for four to six weeks before use.

INGREDIENTS

16 oz (400 g) coconut oil

8 oz (200 g) shea butter

8 oz (200 g) rapeseed oil

8 oz (200 g) olive oil

13 oz (330 g) spring or distilled water

6 oz (147 g) sodium hydroxide

1½ oz (30 g) each of 3 different colors of argile clay

ESSENTIAL OILS

½ tsp (5 ml) lavender

½ tsp (2 ml) cypress

¼ tsp (1 ml) sage

¼ tsp (1 ml) peppermint

Superfatted at 3%

VARIATION

For a deep, heady fragrance, try ylang-ylang, lavender, and vetiver in equal quantities. For something a little spicier, try ylang-ylang and cinnamon

PACKAGING IDEA

Make stripey, colored bands out of deckchair canvas, cut the short ends with pinking shears, wrap these around your soap, and glue into place.

Savon de la Ste Sabine

Traditionally, the soap mixture for making Marseille soaps is boiled for many days, salted, and then the soaps are dried for several months. To simplify this, I have used the double boiler, hot process (HP) method. One great advantage with HP soaps is that they are mild enough to use right away. The four week curing process is to allow the water to evaporate, thus creating a hard, long-lasting bar. You need to use 35 percent water in your formula as opposed to the 30–33 percent used in the cold process.

INGREDIENTS

20 oz (500 g) coconut oil

20 oz (500 g) olive oil

14 oz (350 g) spring or distilled water

6¼ oz (156 g) sodium hydroxide

ESSENTIAL OILS

1 Tbsp (15 ml) any oils of your choosing

Superfatted at 2%

SPECIAL EQUIPMENT

2 stainless steel saucepans (one needs to fit inside the other to form a bain-marie. The smaller pan should have a lid)

Mold (traditionally, the finished soap should be a cube measuring approximately 3½ in/9 cm, so you need a mold that is at least 3½ in/9 cm high. This batch will make 2 soaps)

Sharp knife or cheesewire

1 Weigh the coconut and olive oils and put in the smaller stainless steel saucepan on a low heat to melt. Remove from heat as soon as melting point is reached.

2 Put on your rubber gloves and goggles. Weigh the water in a plastic bucket. Weigh the sodium hydroxide in a jug. Measure out the essential oils.

3 In a well-ventilated area and while wearing gloves, mask, and goggles, pour the sodium hydroxide into the water, stir with a long-handled, stainless steel spoon until dissolved, and then set aside. Remove mask.

4 Put a small bowl or wire rack inside the larger pot so that the smaller pot will not touch the bottom when placed inside (this is to prevent the soap from burning). Place the smaller pot in the larger one and fill the larger pot with water to the level of the oils. Remove the smaller pot and put the larger pot containing water on the heat to boil. You want to maintain a good simmer throughout the cooking process, so you will need to keep topping off the outer pot with water.

5 Pour the caustic solution into the oils. Stir manually for three minutes and then bring the soap to a medium trace using a stick blender. Place the pot of soap inside the larger pot (on the heat). Cover the soap pot with a lid. You should now leave your soap to cook for one hour, topping off the water if necessary.

6 Meanwhile, line your molds with paper (HP soap is stickier than CP, so it is difficult to get out of the mold unless you line it). The hot process takes between one and two hours depending on the formula and the size of the batch. It is ready for testing when the mixture has yellowed and become translucent like petroleum jelly.

7 Test the soap by taking a half teaspoonful from the pot and rolling it into a ball. If it holds together and isn't too sticky, your soap is ready. If not, it needs cooking for longer.

8 When the soap is cooked, remove it from the heat and add the essential oils, stirring in thoroughly. This is also the point where you should add colorant should you wish to. Pour the soap into the mold.

9 Leave the soap to set for approximately 48 hours or until the soap has reached a "hard cheese" consistency. Turn out of the molds and cut into blocks. You can make your soap look truly authentic by stamping it using woodblock carvings.

VARIATION

Include additives such as lavender buds in your soap at trace. Add natural colorants to your oils—alkanet for a purple-gray or annatto for sunny yellow.

PACKAGING IDEA

Wooden boxes designed for displaying wine bottles are great for packaging these soaps and also for using as molds.

Laurus Nobilis

This is my version of Savon d'Alep, a therapeutic olive oil soap made in the Syrian city of Aleppo. Its base is pure pomace olive oil and up to 30 percent laurel (bay) berry oil, which is said to ease arthritis and rheumatism. The oil has a wonderful natural scent, so no additional fragrance is needed. I have used a regular hot process for my soap and in doing so have produced a mild, fragrant, olive-colored castile soap that is marbled through with white.

1 Weigh the olive and laurel oils and put in the smaller stainless steel saucepan.

2 Put on your rubber gloves and goggles. Weigh the water in a plastic bucket. Weigh the sodium hydroxide in a jug.

3 In a well-ventilated area and while wearing gloves, mask, and goggles, pour the sodium hydroxide into the water, stir with a long-handled, stainless steel spoon until dissolved, and then set aside. Remove mask.

4 Put a small bowl or wire rack inside the larger pot so that the smaller pot will not touch the bottom when placed inside. This is to prevent the soap burning. Place the smaller pot in the larger one and fill the larger pot with water to the level of the oils. Remove the smaller pot and put the larger pot containing water on the heat to boil. You want to maintain a good simmer throughout the cooking process, so you will need to keep topping this pot up with water.

5 Pour the caustic solution into the oils. Stir manually for three minutes and then bring the soap to a medium trace using a stick blender. Place the pot of soap inside the larger pot (on the heat) and cover the inner pot with a lid.

6 You should now leave your soap to cook for 2½–3 hours, topping up the water in the large pot as it evaporates. When cooked, the soap will be green and almost translucent, although you may still have some soap that looks like mashed potatoes in there. While you are waiting, line your molds with paper.

7 Test the soap by dipping a teaspoon in the soap pot and putting your tongue against it. If your tongue tingles, you need to continue cooking. If the soap is ready, it will taste of surprisingly fragrant, pure soap. When the soap is cooked, remove it from the heat and simply tip it into your mold, banging the mold down to ensure an even spread.

8 Leave the soap to set for approximately 48 hours or until the soap has reached a "hard cheese" consistency. Turn out of the mold and cut into blocks. Leave the soap to dry for at least two months. It may appear shrunken as the water evaporates, but in the case of Savon d'Alep, that's a good thing!

INGREDIENTS

30 oz (760 g) olive oil

9½ oz (240 g) laurel (bay) berry oil

14 oz (350 g) spring or distilled water

5¼ oz (131 g) sodium hydroxide

14 oz (350 g) water

Superfatted at 2%

SPECIAL EQUIPMENT

2 stainless steel saucepans (one needs to fit inside the other to form a bain-marie. The smaller pan should have a lid)

Mold (traditionally the finished soap should be a cube measuring approximately 3½ in/9 cm, so you need a mold that is at least 3½ in/9 cm high. This batch will make 2 soaps)

HOME SPA

WITH A BIT OF PREPARATION, IT'S POSSIBLE to create a spa environment in your home—all you need is a deep, hot bath; lots of heavenly scented candles; some of your own homemade bath salts, scrubs, and oils; and the perfect soap for the occasion. Once you climb out of the bath, you can smother yourself in your own homemade body butter (see page 133), and what's more, you will know for certain that the ingredients in your products are the best that money can buy and that you have not fallen into the trap of spending fortunes on a brand name.

The soaps in this section will detoxify, rejuvenate, and stimulate your skin. They contain natural minerals such as Dead Sea mud and sea salt, along with pure, soothing botanicals, including aloe vera gel. All of these products, along with natural waxes, butters, ,and plant extracts are widely available through soapmaking suppliers, and each will add that special "spa experience" to your soap.

Seaweed and Aloe Vera

Seaweed is packed with vitamins and minerals and is a great additive for your soap. If you replace some of the water content with aloe vera juice or gel, the soap will soothe even the most sensitive of skins.

1 Weigh the coconut oil, cut into small chunks, and put in a plastic bucket. Weigh and add the olive and hempseed oils.

2 Put on your rubber gloves and goggles. Weigh the water in a plastic bucket. Weigh the sodium hydroxide in a jug. Weigh the kelp and dilute the oxide. Weigh the aloe vera. Measure out the essential oils and pour over the kelp.

3 In a well-ventilated area and while wearing gloves, mask, and goggles, pour the sodium hydroxide into the water, stir with a long-handled, stainless steel spoon until dissolved, and then set aside. Remove mask.

4 Pour the caustic solution into the oils. Stir manually until the fats and oils have dissolved and then bring the soap to a light trace using a stick blender. Add the kelp, the aloe vera, and the colorant and stir thoroughly.

5 Pour into a 3 lb (1½ kg) prepared mold. Leave the soap to set for approximately 48 hours or until it has reached a "hard cheese" consistency. Turn out of the mold and cut into bars. Cure the bars for four to six weeks before use.

INGREDIENTS

16 oz (400 g) coconut oil

16 oz (400 g) olive oil

8 oz (200 g) hempseed oil

13 oz (330 g) spring or distilled water

6 oz (150 g) sodium hydroxide

2 oz (50 g) powdered kelp

1 Tbsp (15 ml) diluted green oxide

4 oz (100 g) aloe vera gel

ESSENTIAL OILS

2 tsp (10 ml) rosemary

1 tsp (5 ml) lavender

Superfatted at 3%

VARIATION

Replace the kelp with spirulina powder and use an essential oil blend of 2 tsp (10 ml) ylang-ylang, ½ tsp (2½ ml) lemon, and ½ tsp (2½ ml) gulab attar.

PACKAGING IDEA

Crisscross hemp twine around the soap and knot. Thread the ends through two seashells and knot the twine to secure.

Sea Salt

Sea salt is a superb natural softener for your skin and, if used undiluted, is also an excellent exfoliant. For this recipe, it is best to use medium to fine grains of salt as the large grains can be scratchy.

1 Weigh the coconut and palm kernel oils and the cocoa butter, cut into small chunks, and put in a plastic bucket. Weigh and add the olive oil.

2 Put on your rubber gloves and goggles. Weigh the water in a plastic bucket. Weigh the sodium hydroxide in a jug. Weigh the sea salt and dilute the oxide. Measure out the essential oils and pour them onto the sea salt.

3 In a well-ventilated area and while wearing gloves, mask, and goggles, pour the sodium hydroxide into the water, stir with a long-handled, stainless steel spoon until dissolved, and then set aside. Remove mask.

4 Pour the caustic solution into the oils. Stir manually until the fats and oils have dissolved and then bring the soap to a light trace using a stick blender. Add the colorant and stir thoroughly.

5 When the soap is at medium trace, stir in the salt/essential oil blend and pour into your prepared 3 lb (1½ kg) mold. Leave the soaps to set for approximately 48 hours or until the soap has reached a "hard cheese" consistency. Turn out of the molds and cut into bars. Cure the bars for four to six weeks before use.

INGREDIENTS

16 oz (400 g) coconut oil

12 oz (300 g) palm kernel oil

4 oz (100 g) cocoa butter

8 oz (200 g) olive oil

13 oz (330 g) spring or distilled water

5½ oz (157 g) sodium hydroxide

16 oz (400 g) fine or medium sea salt

15 ml (1 Tbsp) diluted blue oxide

ESSENTIAL OILS

2 tsp (10 ml) lemon tea tree

1 tsp (5 ml) sweet orange

Superfatted at 4%

VARIATIONS

Leave out the coloring and replace half of the salt with powdered kelp to make a green seaweed and sea salt combination.

PACKAGING IDEA

Wrap the soap in pale blue and white striped fabric or wallpaper and package together with a jar of Shea Butter Whip (see page 133).

Dead Sea Mud

Dead Sea mud stimulates blood circulation, hydrates the skin, and is reputed to have antiaging qualities. Try it on the roots of your hair because it is known to strengthen them, and you could make a face mask with any leftover mud.

INGREDIENTS

16 oz (400 g) coconut oil

2 oz (50 g) shea butter

4 oz (100 g) olive oil

10 oz (250 g) sunflower oil

10 oz (264 g) spring or distilled water

5 oz (124 g) sodium hydroxide

2 tsp (10 ml) vitamin E oil

4 oz (100 g) Dead Sea mud

ESSENTIAL OILS

2 tsp (10 ml) rosemary

1 tsp (5 ml) lavender

Superfatted at 3%

1 Weigh the coconut oil and shea butter, cut into small chunks, and put in a plastic bucket. Weigh and add the olive and sunflower oils.

2 Put on your rubber gloves and goggles. Weigh the water in a plastic bucket. Weigh the sodium hydroxide in a jug. Weigh the mud. Measure out the essential oils and mix them with the mud.

3 In a well-ventilated area and while wearing gloves, mask, and goggles, pour the sodium hydroxide into the water, stir with a long-handled, stainless steel spoon until dissolved, and then set aside. Remove mask.

4 Pour the caustic solution into the oils. Stir manually until the fats and oils have dissolved and then bring the soap to a light trace using a stick blender. Add the vitamin E oil and the mud and stir to medium trace.

5 Pour the soap into a 3 lb (1½ kg) prepared mold. Leave the soap to set for approximately 48 hours or until it has reached a "hard cheese" consistency. Turn out of the mold and cut into bars. Cure the bars for four to six weeks before use.

VARIATION

Add ¼ oz (5 g) of ground loofah as an exfoliant or pour the soap directly into a loofah and cut into slices when set.

PACKAGING IDEA

Package this soap in a drawstring bag together with the Shea Butter Whip (see page 133), a Bath and Body Oil (see page 137), and a slice of loofah.

OUT OF AFRICA

AFRICA OFFERS A WEALTH OF BOTANICAL wonders that can be included in soap recipes. Many of these plants, herbs, and butters have been used for centuries by the local people to heal skin disorders and to protect the skin from the blistering sun.

The marula tree has been a source of nutrition in Africa since 10,000 BC. In South Africa and Zimbabwe, the oil, extracted from the fruit, is used as a skin cleanser and a baby oil. The oil produced from the seeds of the baobab (or "upside down") tree is rich in vitamins A and F—excellent for damaged skin—and baobab powder can be used as a natural exfoliant.

Shea butter has incredible moisturizing benefits, and soaps and creams made with cocoa butter have a wonderful richness. The fair trade movement has been a life-changing force to the communities of mainly women who make their livelihoods from harvesting and processing these butters—particularly in Ghana and Tanzania—and our use of these exotic ingredients in our cosmetics means so much to Africa.

Baobab and Marula

The seeds of the Baobab tree (much loved by elephants) are full of edible goodness, and the oil is superbly moisturizing and gentle in cosmetic products. Marula oil, extracted from the kernels, has an ideal composition of vitamins, minerals, and fatty acids.

INGREDIENTS

20 oz (500 g) coconut oil

2 oz (50 g) shea butter

8 oz (200 g) sunflower oil

4 oz (100 g) baobab oil

6 oz (150 g) marula oil

13 oz (330 g) spring or distilled water

6 oz (155 g) sodium hydroxide

ESSENTIAL OILS

1½ tsp (7½ ml) rose geranium

1 tsp (5 ml) lavandin

Superfatted at 3%

1 Weigh the coconut oil and shea butter, cut into small chunks, and put in a plastic bucket. Weigh and add the sunflower, baobab, and marula oils.

2 Put on your rubber gloves and goggles. Weigh the water in a separate plastic bucket. Weigh the sodium hydroxide in a jug. Blend the essential oils.

3 In a well-ventilated area and while wearing gloves, mask, and goggles, pour the sodium hydroxide into the water, stir with a long-handled, stainless steel spoon until dissolved, and then set aside. Remove mask.

4 Pour the caustic solution into the oils. Stir manually until the fats and oils have dissolved and then bring the soap to a light trace using a stick blender.

5 Pour the soap into a 3 lb (1½ kg) mold. Leave the soap to set for approximately 48 hours or until it has reached a "hard cheese" consistency. Turn out of the mold and cut into bars. Cure the bars for four to six weeks before use.

VARIATION

In Africa soap is used as much for its healing as its cleansing properties and 2 tsp (10 ml) of neem oil added to this recipe will help to heal cuts.

PACKAGING IDEA

Banana leaves and corn husks both make great packaging materials either nude from the tree or pulped and made into handmade paper.

Rooibos and Wild Honey

Rooibos is a herb that is unique to South Africa and sold worldwide as a tea. For the soapmaker, it offers antioxidant qualities. Honey is often used to help acne sufferers while beeswax gives your soap a hard waxy texture and makes it last longer.

1 Weigh the water and bring to the boil. Mix with the rooibos and the honey to make a tea. Leave to go cold. Sieve out the tea leaves and set them aside. Weigh the olive and palm kernel oils and the beeswax, and put them in a stainless steel saucepan on a low heat to melt. Remove from heat as soon as melting point is reached.

2 Put on your rubber gloves and goggles. Weigh the sodium hydroxide in a jug. Blend the essential oils.

3 In a well-ventilated area and while wearing gloves, mask, and goggles, pour the sodium hydroxide into the rooibos and honey tea, stir with a long-handled, stainless steel spoon until dissolved, and then set aside. Remove mask.

4 Check the temperature of both the caustic solution and the oils, and when both reach approximately the same temperature, pour the caustic solution into the oils. Stir manually for three minutes and then bring the soap to a medium trace using a stick blender.

5 Add the essential oils and stir. If you want to add a handful of the brewed tea leaves, this is the time to throw them into the pot. Stir until evenly distributed, then pour the soap into a mold.

6 Leave the soap to set for approximately 48 hours or until it has reached a "hard cheese" consistency. Turn out of the mold and cut into bars. Cure the bars for four to six weeks before use.

INGREDIENTS

13 oz (330 g) spring or distilled water

2 oz (50 g) rooiboos tea

4 oz (100 g) honey

20 oz (500 g) olive oil

16 oz (400 g) palm kernel oil

4 oz (100 g) beeswax

5 oz (135 g) sodium hydroxide

ESSENTIAL OILS

2 tsp (10 ml) rosemary

½ tsp (2.5 ml) lavandin

½ tsp (2.5 ml) rose geranium

Superfatted at 3%

VARIATIONS

Honeybush tea can be substituted for the rooibos—try adding a blend of sweet orange, ylang-ylang, and a couple of drops of ginger and clove oil.

PACKAGING IDEA

Make use of brightly colored African beads. Thread them on a leather thong or linen twine and tie around your soap. Present on a wooden soap dish.

Melon and Shea Butter

Watermelon seed oil comes from the Kalahari melon grown in dry, rural regions of Africa, particularly in northern Namibia. Coupled with unrefined shea butter and coconut oil, it provides a highly moisturizing soap with a good, rich lather.

INGREDIENTS

16 oz (400 g) coconut oil

12 oz (300 g) shea butter

12 oz (300 g) watermelon seed oil

13 oz (330 g) spring or distilled water

5 oz (147 g) sodium hydroxide

ESSENTIAL OILS

½ tsp (2½ ml) tea tree

½ tsp (2½ ml) lavandin

½ tsp (2½ ml) rosemary

1½ tsp (7½ ml) sweet orange

Superfatted at 3%

1 Weigh the coconut oil and shea butter, cut into small chunks, and put in a plastic bucket. Weigh and add the watermelon oil.

2 Put on your rubber gloves and goggles. Weigh the water in a plastic bucket. Weigh the sodium hydroxide in a jug. Blend the essential oils.

3 In a well-ventilated area and while wearing gloves, mask, and goggles, pour the sodium hydroxide into the water, stir with a long-handled, stainless steel spoon until dissolved, and then set aside. Remove mask.

4 Pour the caustic solution into the oils. Stir manually until the fats and oils have dissolved and then bring the soap to a light trace using a stick blender.

5 Pour the soap into 3 lb (1½ kg) prepared mold. Leave the soap to set for approximately 48 hours or until it has reached a "hard cheese" consistency. Turn out of the mold and cut into bars. Cure the bars for four to six weeks before use.

VARIATION

For a heady, flowery scent, use equal quantities of rose geranium, lavendin, and ylang-ylang with a couple of drops of Roman chamomile.

PACKAGING IDEA

Try wrapping this soap in banana leaves and using a strip of the leaf as a string to secure it. Corn husks also make interesting packaging.

African Black Soap

No section on Africa would be complete without reference to African black soap, known also as *alata* in Ghana and *ose dudu* in Nigeria. Black soap is produced in many rural communities throughout Africa from heavily-guarded recipes and has been prized for generations for its gentle moisturizing qualities.

The ingredients used in African black soap vary from region to region depending on what grows locally, but most comprise of either an olive-, palm kernel- or coconut-based oil hand-pressed within the villages. Where shea butter (karite) is available, this is often included as a moisturizing additive. The uniqueness of African black soap is the fact that the lye is made by dripping rainwater through the ashes of roasted cocoa pods or plantain skins—not something most of us could try at home.

When lye is made in this fashion, it is difficult to measure the density and the resulting solution is closer to potassium hydroxide (used to make liquid soap) than sodium hydroxide (used for bar soaps) and as a result, true African black soap is often soft and malleable. It is also sometimes diluted and sold as a liquid soap for use as a shampoo. That said, you can find on the market a number of hard bar soaps originating in Africa that describe themselves as "African black soap." These are still a long way from the traditional village soaps that are created with much pride and joy in "soap kettles," where the lye and the fats/oils and water are boiled together over an open fire until the soap raises to the surface and is skimmed off and formed into balls or rough blocks.

FESTIVE SOAPS

SOAPS AND TOILETRIES MAKE GREAT gifts, and for those who plan to run a soapmaking business you will find that you sell more soaps during festive seasons than at any other time of the year. The only problem is the stacks of Christmas pudding or pumpkin shaped soaps you will find filling up your shelves the day after the holiday.

However, there is a simple solution. Make your soaps using traditional Christmas fragrances, such as cinnamon, orange, and clove; use as much pumpkin oil as you like in your Thanksgiving soap; but, with the exception of heart shapes (which boom around St. Valentines day and do well right through the year), stick to oblongs and squares and use only your packaging to express the festivities.

That way, you can repackage spicy smelling soaps after the event to suggest a season-proof oriental theme, and you can promote your pumpkin soap by highlighting the antioxidant elements of pumpkin seed oil (you could even repackage this soap for pregnant moms; pumpkin seed oil is great for stretch marks). Problem solved.

Christmas Spice

The rich scents of cinnamon, orange, and clove are synonymous with Christmas and lovely to have in your bathroom to welcome holiday guests. Be aware that soaps containing clove oil trace very quickly, so have your molds ready before you mix your soap.

INGREDIENTS

16 oz (400 g) coconut oil

12 oz (300 g) olive oil

4 oz (100 g) watermelon oil

8 oz (200 g) shea butter

13 oz (330 g) spring or distilled water

5 oz (147 g) sodium hydroxide

4 oz (100 g) cocoa powder

dried orange slice or cinnamon stick, to decorate (optional)

ESSENTIAL OILS

2 tsp (10 ml) sweet orange

½ tsp (2.5 ml) cinnamon

½ tsp (2.5 ml) clove

Superfatted at 4%

1. Weigh the coconut oil and chop into small chunks. Weigh the olive and watermelon oils and the shea butter and put in a plastic bucket.

2. Put on your rubber gloves and goggles. Weigh the water in a plastic bucket. Weigh the sodium hydroxide in a jug.

3. Measure out the essential oils and blend together. Measure out the cocoa powder. Make a paste by adding around 1 tablespoon of water. Add the essential oil blend to the cocoa paste.

4. In a well-ventilated area and while wearing gloves, mask, and goggles, pour the sodium hydroxide into the water, stir with a long-handled, stainless steel spoon until dissolved, and then set aside. Remove mask.

5. Pour the caustic solution into the oils. Stir manually until the fats and oils have dissolved and then bring the soap to a light trace using a stick blender. Add the cocoa powder and stir thoroughly.

6. Pour the soap into a 3 lb (1½ kg) mold. Leave the soap to set for approximately 48 hours or until it has reached a "hard cheese" consistency. Turn out of the mold and cut into bars.

7. Press the dried orange or the cinnamon stick into the surface of your soap, if using. If it won't stick, wet the surface of the soap slightly. Cure the bars for four to six weeks before use.

VARIATION

Replace the cinnamon and clove essential oils with the same quantities of frankincense and myrrh essential oils.

PACKAGING IDEA

Instead of cutting the soap into bars, roll into balls (wearing rubber gloves), tie a ribbon around, and stick cloves into the surface to form a pomander.

Bollywood

During a particularly creative week, I dressed my soaps as Indian princesses in fine silk fabric and jewelled bindis. Harrods of London made an order for a range of similarly attired soaps, lotions, and balms, and this is the soap I created for them.

1 Weigh the coconut and palm oils and the shea butter, chop into small chunks, and put in a plastic bucket. Weigh and add the olive oil.

2 Put on your rubber gloves and goggles. Weigh the water in a plastic bucket. Weigh the sodium hydroxide in a jug. Measure out the essential oils and blend them together. Dilute the colorant.

3 In a well-ventilated area and while wearing gloves, mask, and goggles, pour the sodium hydroxide into the water, stir with a long-handled, stainless steel spoon until dissolved, and then set aside. Remove mask.

4 Pour the caustic solution into the oils. Stir manually until the fats and oils have dissolved and then bring the soap to a light trace using a stick blender. Add the colorant and the essential oil blend and stir thoroughly

5 Pour the soap into a 3 lb (1½ kg) mold. Leave the soap to set for approximately 48 hours or until it has reached a "hard cheese" consistency. Turn out of the mold and cut into bars. Stick the bindis into place. Cure the bars for four to six weeks before use.

INGREDIENTS

16 oz (400 g) coconut oil

4 oz (100 g) palm oil

8 oz (200 g) shea butter

12 oz (300 g) olive oil

13 oz (330 g) spring or distilled water

6 oz (148 g) sodium hydroxide

ESSENTIAL OILS

1 tsp (5 ml) rose geranium

½ tsp (2.5 ml) frankincense

½ tsp (2 ml) ylang-ylang

¼ tsp (1.5 ml) mandarin

¼ tsp (1 ml) cinnamon

¼ tsp (1 ml) bergamot

¼ tsp (1 ml) lemon

¼ tsp (1 ml) gulab attar

COLORANT

D&C colorant or oxide of your choice—dilute ¼ tsp (1 ml) D&C colorant in 4 tsp (20 ml) water or 1 Tbsp (5 ml) oxide in 6 tsp (30 mls) water and add drop by drop at trace until you achieve the desired color

Superfatted at 4%

VARIATION

Just after cutting the soap, sprinkle a metallic mica powder onto the surface and substitute the fabric for gold or silver paper.

PACKAGING IDEA

Package this soap together with a Bath and Body Oil (see page 137) and a Bath Melter (see page 134) in a foil box with an acetate lid.

I Do

Weddings are big business. Offering to make favors is a great way of showing off
your soapmaking skills and building your client list. In this recipe, I have
used pure silk fabric since it really does make the soap feel silky.

INGREDIENTS

16 oz (400 g) coconut oil

12 oz (300 g) olive oil

4 oz (100 g) hempseed oil

4 oz (100 g) sweet almond oil

4 oz (100 g) cocoa
butter, grated

13 oz (330 g) spring or
distilled water

6 oz (149 g) sodium hydroxide

Gold mica (optional)

ESSENTIAL OILS

2 tsp (10 ml) sweet orange

½ tsp (2.5 ml) geranium

½ tsp (2.5 ml) palmarosa

COLORANT

D&C colorant or oxide of
your choice—dilute ¼ tsp (1 ml)
D&C colorant in 4 tsp (20 ml)
water or 1 Tbsp (5 ml) oxide in
6 tsp (30 mls) water and add
drop by drop at trace until you
achieve a pastel tint

SPECIAL EQUIPMENT

8 in (20 cm) square of
undyed silk

fancy silicone ice cube trays

Superfatted at 4%

1 Weigh the coconut oil, cut into small chunks, and put in a plastic
bucket. Weigh and add the olive, hempseed, sweet almond oils, and the
cocoa butter.

2 Weigh the water in a plastic bucket. Weigh the sodium hydroxide in
a jug. Measure out the essential oils and blend together. Dilute the
colorant

3 In a well-ventilated area and while wearing gloves, mask, and goggles,
pour the sodium hydroxide into the water, stir with a long-handled,
stainless steel spoon until dissolved. Add the piece of silk to the lye
solution and stir until dissolved, and then set aside. Remove mask.

4 Pour the caustic solution into the oils. Stir manually until the fats and
oils have dissolved and then bring the soap to a light trace using a stick
blender. Add the essential oil blend and stir thoroughly. Now add the
colorant drop by drop until you achieve the desired color.

5 Pour the soap into silicone ice cube trays. Leave the soaps to set for
approximately 48 hours or until they have reached a "hard cheese"
consistency. Turn out of the mold. Sprinkle powdered gold mica onto
the surface of the soaps. Cure the soaps for four to six weeks before use.

VARIATION

Split the batch into two jugs at trace and add blue colorant to one
jug and pink to the second. Pour into two separate molds.

PACKAGING IDEA

Wrap the soaps in silk paper and tie with silk ribbon
or put three or four in a small voile bag.

SPECIAL EFFECTS

SPECIAL EFFECTS CAN BE ACHIEVED through changes and combinations of color, shape, presentation, and purpose, and by using any one of the thousands of decorative items available at craft stores as well as on beaches, in forests, and at food stores.

COLOR

There are both natural and artificial ways to color your soap (see pages 18–19). The color of your base oils will affect the final color, but the best way to experiment is to make a plain batch of soap and to divide it at light trace into several jugs, coloring each one with a different oxide, pigment, spice or infused oil. The following recipe will give you a neutral-colored soap base to play with:

16 oz (400 g) coconut oil
12 oz (300 g) palm oil
12 oz (300 g) sunflower oil
13 oz (330 g) water
6 oz (152 g) sodium hydroxide

SWIRLS

To achieve a good swirl in your soap, divide the batch into three parts base color and one part contrasting color. Pour a layer of base into your mold and then pour lines of your contrasting color over the top. Swirl the soap together by running a skewer or a fork through the colored lines. Add another layer of the base and more lines of color and swirl again. Repeat until all the soap is used up. Good swirls come from plenty of practice and different effects can be achieved by "pulling" the color in different directions.

LAYERS

Using a loaf or box mold you can make endless layers of different colors by pouring soaps directly on top of each other. It is important to ensure that each layer is firm enough to support the layer on top, so your bottom layer needs to have reached a very heavy trace before the next layer is added. One way to achieve this is to divide and color your soaps at a very light trace. Using a stick blender, zap the jug containing the soap you want to use for your base layer until it is very thick and then scrape it into your mold. By this time, your second jug of soap will have begun to thicken, and you can pour this directly on top. If you are planning on building three layers, zap the contents of the second jug as well so that the soap will be thick enough to support your third layer.

LANDSCAPE SOAP

Wonderful landscape and seascape effects can be achieved by pouring random amounts of different colored or textured soaps into a loaf mold at heavy trace. To create a seascape, start with a yellow or neutral colored soap and add some oatmeal for texture. Pour this into the mold. Now pour some blues or greens (or both) in lines and blobs over the oatmeal layer until the soap is about 5 cm (2 in) deep in total. Your next layer is for the sky and could be made with pale blue or grey (alkanet) soap with random blobs of white soap (for clouds), and perhaps some pink lines to give a sunrise effect. You also want a thick line of yellow or red soap poured along one long edge of the mold to form the sun. When planning your design you need to think about how a stick of rock candy works. Patterns are formed in long strips that run right through the rock—when placing your colors remember that the soap will be sliced and so you need the color to run right through the mold.

The best landscape soaps are made by accident. On soapmaking day, keep an empty loaf mold on hand and fill it with the scrapings from each batch of soap you make, positioning them in random blobs or thin layers as you see fit. You can keep adding to the mold as you make different colored soaps, and you will be amazed at the beautiful effects you can achieve. You can keep a mold going in this way for up to a week—any longer, and the different soap layers may not hold together well because the soap in the mold will be too hard. Remember that if you are planning to sell your landscape soaps, you need to work with just one base formula and ensure that all the colorants/additives you use are on your assessment (see pages 138–141) and listed on your labels.

SHAPE

The shape of your soap is determined by the shape of your mold or your cutting device. If you want fancy shapes the best results will be achieved using flexible silicone or latex molds and these are made not only for soap making but for garden ornaments, architectural features, candles, model making and numerous other arts and crafts.

The cupid plaque in the illustration was made using a mold intended for a concrete slab. I thought it would be fun to create a soap plaque that you could hang in the shower, the idea being

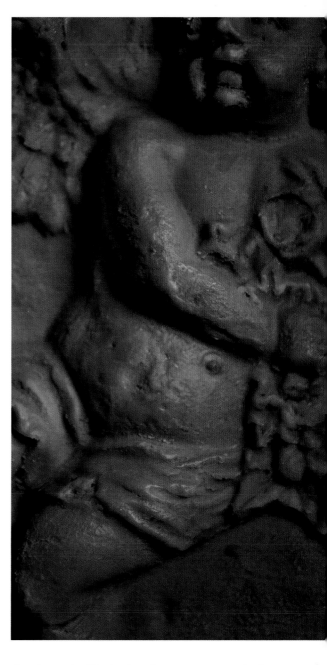

that instead of bringing the soap to your body, you fill your sponge with soap by rubbing it on the soap plaque. Once the soap is hard, you need only to drill a couple of holes in it, and you can hang it in the shower on a cord. You could also display it in a box frame you make; then you can refill this with a fresh soap plaque when your soap "fixture" begins to wear too thin.

Liquid Soap

THE BASICS

IF YOU ARE PLANNING TO START YOUR own soapmaking business, it is worth mastering the art of liquid soapmaking because liquid soaps outsell bar soaps eight to one. There are several ways to make liquid soap, but the main method used in this book is the paste method. Using a double boiler, oils are saponified with potassium hydroxide and cooked together for several hours until they form a glutinous soap gel. The gel is then diluted with water and essential oils and colorants are added. Compared to bar soapmaking, this process is quite long-winded but the advantages are first that the soap paste can be stored in a cool place for up to a year before dilution, and second the soap is ready to use as soon as it is cooked.

Any product containing water can attract bacteria, and for this reason, you will find that the pH value of most natural liquid soaps is relatively high at around 10.3, and considerably higher if the soaps do not contain a preservative. A high pH helps reduce the possibility of bacteria forming, and while some essential oils will help to preserve your product, there is, at the time of writing, no truly effective natural preservative available. Due to the high pH, a natural liquid soap is probably more suitable as a hand soap than a face soap.

The second method involves cooking oils, potassium hydroxide, and alcohol together. While the alcohol speeds up the saponification process, this is a tricky method to use at home because alcohol and kitchen stoves (especially gas ones) are a potential recipe for disaster.

The third and speediest method of making liquid soap is to dissolve cold or hot processed soaps in boiling water. Depending how much water you use, you can achieve a relatively thick liquid soap, but it does have a tendency to form a skin, and you cannot make a clear soap using this simple method.

FATS AND OILS (ACIDS)

The choice of oils and fats is infinite, but the quality of the soap you produce will depend on this choice. When formulating a liquid soap, your main consideration is to create a mild, clear soap with rich bubbles. For both clarity and lather, coconut oil is a good choice. It has a large proportion of lauric acid, which is highly soluble and as such will contribute greatly to the clarity of the soap and to the quality of the lather. However, a soap containing 100 percent coconut oil can be drying on the skin so this needs to be balanced with other oils.

On pages 10–14, you will find a full description of the properties of the oils most commonly used in soap and cosmetics, and from this list you can select other oils for your formula, depending on what you want to achieve. Both olive and sunflower oils are great for bulking out the formula and are both mild and emollient. Castor oil will give you a good foam and is a great choice for both a shaving soap and a shampoo, but for gentleness you could include (in small quantities) moisturizing oils such as avocado, apricot kernel, and shea butter.

POTASSIUM HYDROXIDE (ALKALI)

It is important that the correct amount of alkali is used to saponify the oils you have selected. If you look at the saponifcation chart on page 25, you will see two sets of numbers (SAP values) next to each oil. When making liquid soap, we use potassium hydroxide (KOH) as opposed to sodium hydroxide

(NaOH), so the KOH values are the ones to use. See page 24 for instructions on multiplying the weight of oil or fat by KOH.

MAKING LIQUID SOAP

Liquid soap is made using a hot process method. As you are using a powerful alkali to saponify your oils, it is important that you treat the alkali with respect and follow all the safety measures described on page 23. Remember to always wear rubber gloves and eye protection when working with lye or fresh soap and wear a face mask when pouring the lye into the water. Take care to accurately weigh out all fats and oils and water and lye. For liquid soap, calculate 35 percent of the weight of the oils and use that amount of water in the lye solution. You also have the option of replacing some of that water content with aloe vera gel.

The lye water is then added to the oils and mixed with a stick blender until the mixture has emulsified (traced). This process takes considerably longer using potassium hydroxide than sodium hydroxide so don't panic if the mixture doesn't seem to be tracing, just keep going. Once the mixture is creamy, put the soap pot into a larger pot containing water (bain-marie), and then place on the heat to cook for around three hours at a rolling boil.

POTASSIUM CARBONATE

During the cooking process, the soap goes through a number of stages. For the first 30 minutes, the mixture will appear to separate and even curdle, but keep stirring every ten minutes until it comes together. Once this happens, the soap will begin to resemble apple sauce, quickly followed by mashed potato, and then it turns to thick toffee which can be almost impossible to stir. After a period of around three hours (depending on your batch size), the sticky toffee will turn translucent and resemble deep yellow petroleum jelly. At this point, your soap is ready.

The sticky toffee stage is manageable with small batches of soap, but if you are making a large batch, you could find yourself neck deep in a gluey mess trying to stir the soap at the bottom of the pot. To avoid this situation, I use potassium carbonate in all my liquid soap formulas. While it is also an alkali, it has the magic property of keeping your soap mixable right through the process. This is a real lifesaver for large batches, but the downside is that if you mix alkalis together, it takes longer to reach trace, and separation is more likely to occur. Once again, just keep stirring and the mixture will blend together.

To calculate how much potassium carbonate to use, simply add together the weight of the oils and add carbonate at 2 percent of that weight.

SUPERFATTING

It is not advisable to superfat a liquid soap because free oils plus water can equal bacteria. However, potassium carbonate is an alkali, and to ensure your soap remains mild it is necessary to make an adjustment for it when working out how much potassium hydroxide you need in your formula. If you are working out the amount of potassium hydroxide you need from the chart on page 25, simply deduct 2 percent from the total. Alternatively, if you are using an online SAP calculator (see Internet Resources on page 143), set the superfatting option to 2 percent.

TESTING

Once your soap has turned translucent, you should test it for clarity and mildness. To do this, take a small spoonful of soap from the pot and weigh it, then put it in a glass or plastic tumbler. You should now dilute this in twice its weight of boiling water (for example, for ½ oz/14 g soap paste, add 1 oz/28 g boiling water). If the soap is ready, the dissolved solution should be clear. You can also test it for mildness using pH papers. If the soap is ready, the paper will indicate a pH of below 10.5.

DILUTING

The paste you have made is a highly concentrated pure soap, and this can be transformed into a liquid consistency by adding boiled spring or distilled water. My personal preference is 2 parts water to one part soap paste; this produces a stable, cleansing soap that works well in a pump applicator. It is reasonable to assume that if you use less water your soap will be thicker, but in fact, using less water simply gives you a stronger concentration of soap. You can dilute as much or as little of the soap paste as you like, storing any excess in a plastic lidded tub. Simply weigh the amount of paste you want to dilute and add twice the amount of boiling spring water. Do not stir (this will introduce bubbles), just cut the paste with a knife to speed up the process. Soap dilution can take quite a long time, especially for large batches, so it is best to add the boiling water, cover the pot, and leave it overnight. If the soap has not fully dissolved by the morning, heat it gently and cut any remaining paste until the process is complete.

NEUTRALIZING

Once your soap has dissolved, check the pH again. If it is over 10.5, you can neutralize it with a citric acid formula. Dissolve 1 part citric acid in 4 parts water, and, using a pipette, add this solution to your soap until the pH drops to under 10.

FRAGRANCING

Liquid soap can be fragranced with up to 2 percent essential or fragrance oils. Warm the diluted soap to just above room temperature before adding fragrance. At this stage, you can also add a few drops of liquid cosmetic or food colorant but be aware that a little goes a long way. Some essential oils will cloud your soap initially, but this will settle.

PRESERVING

Any product containing water has the potential to attract bacteria, but bacteria is often introduced into products such as creams and lotions by dipping your fingers in the pot. Liquid soaps are usually packaged in sealed pump applicators, which reduce the risk of bacteria entering the product, and there are a number of natural liquid soaps on the market that rely entirely on a pH of around 10 and the use of antibacterial essential oils such as tea tree or rosemary to fight the possibility of bacteria.

The choice is yours, but if you intend to sell your soap, you should consider using one of the non-paraben based preservatives on the market. There are a number of cosmetic preservatives available, but before making a selection, you need to remember that you are working with a high pH, and many preservatives are not designed for this. My preservative of choice is Optiphen, which is a paraben- and formaldehyde-free blend of phenoxyethanol and Capryl glycol. ND is the Optiphen grade needed for liquid soap, and I have found it works well at 1 percent. It is widely available through distributors in the UK and in the USA (see resource section on page 143).

THICKENING

While the quality of your natural liquid soap will be extremely high, it may tend to be thinner than mass-produced liquid detergents. Typical thickening agents such as agar, guar gum, and carrageenan do not work well in products with a high pH and tend to turn liquid soap gloopy. Outside Europe, borax is widely used for both its thickening and chelating properties, but it is restricted under EU legislation as it known to irritate the skin. Some soapmakers use salt as a natural thickener, and you can also replace up to 4 percent of your potassium hydroxide with sodium hydroxide during the paste formation to assist with thickening, although I have not had much success with this.

The best thickener I have is a cellulose gum made from plant fibers. If diluting 2 oz (50 g) of soap paste, sprinkle up to third of a teaspoon cellulose gum into 5 oz (150 g) boiling water and whisk thoroughly. Add this water to your soap paste to dilute it. You can adjust the quantity of gum according to how thick you want your soap. If you intend to put your soap into a foam pump, do not attempt to thicken it because it will clog up the mechanism.

BOTTLING

Once your diluted soap is cool, sterilize your bottles with boiling water, pour the soap into the bottle, and seal.

EQUIPMENT

Double boiler (Bain-Marie)
Two stainless steel pots, one to fit inside the other so water can be boiled in the outer pot. The smaller pot needs a tight-fitting lid.
Kitchen scales weighing in small increments. You need specialist scales if you are making soap to sell.
Long-handled, stainless steel spoon
Knife to cut fats
Stick blender
Plastic bucket
Plastic jugs x 2
Rubber spatula
Plastic lidded container to store the paste in
Pump bottles for the finished soap
Rubber gloves
Eye goggles
Dust mask
Vinegar

INGREDIENTS

Oils and fats of your choice
Potassium hydroxide
Potassium carbonate (optional)
Spring or distilled water
Citric acid
Preservative (optional)

COOKING LIQUID SOAP

1. Melt the oils and add the lye mixture. Stir until the mixture forms an emulsion and then put in a bain-marie.

2. The soap will go through an "apple sauce" stage.

3. Eventually the soap will become translucent and resemble yellow petroleum jelly. Test the soap for clarity and mildness and then dilute.

> ### SAFETY
> Potassium hydroxide and potassium carbonate are both alkalis that can cause burns if not handled with care. Please read and follow all the safety instructions on page 23.

Basic Liquid Soap Paste

This recipe will produce a gentle, clear, all-purpose soap. You can add whatever essential oils you like, perhaps tea tree for a kitchen soap or a geranium and lavender combination for something a little more feminine.

1 Weigh the coconut, sunflower, and olive oils and put them in a stainless steel saucepan. Weigh the water and pour into a plastic bucket. Half fill the larger saucepan with tap water and put it on the heat.

2 Put on your gloves, goggles, and mask. Weigh the potassium hydroxide and potassium carbonate separately and then combine. Add the potassium hydroxide and potassium carbonate mixture to the water and stir for two minutes.

3 Melt the oils on a low heat and stir. Remove the soap pot from the heat. Still wearing your protective kit, pour the alkali mixture into the pot and stir again. Bring the soap to trace using a stick blender. (The oils and liquids may separate—keep mixing until they come together again.)

4 Place the soap pot into the larger pot and bring the water to a rolling boil. Cover the pot. Top off water in outer pot as it evaporates. Stir every 15 minutes for approximately three hours or until the mixture in the pot resembles deep yellow petroleum jelly.

5 Remove from the heat and test (see page 95). If the soap is ready, store in a plastic lidded container or dilute as per the instructions below. If not, continue to cook until it tests clear.

DILUTION
To fill a 8 fl oz (250 ml) container, weigh 3 oz (83 g) of soap paste and dilute with 6 oz (166 ml) boiling spring water. Once dissolved, heat until tepid and add up to ½ tsp (2.5 ml) essential oil of your choice. Also add preservative at this stage, should you require it, following the manufacturer's recommended ratio. Allow to cool and then pour into a sterilized bottle.

INGREDIENTS FOR UNDILUTED SOAP PASTE

8 oz (200 g) coconut oil

8 oz (200 g) sunflower oil

4 oz (100 g) olive oil

7 oz (180 g) spring water

4 oz (108 g) potassium hydroxide

⅓ oz (10 g) potassium carbonate

Shaving Mousse

Light, foaming mousses are created by putting your liquid soap into a foam pump applicator. This formula is smooth, moisturizing, and rich in bubbles. Fragrance with cedarwood for men and geranium and mandarin for a floral feminine scent.

INGREDIENTS FOR UNDILUTED SOAP PASTE

12 oz (300 g) coconut oil
4 oz (100 g) castor oil
4 oz (100 g) olive oil
7 oz (180 g) spring water
4½ oz (112 g) potassium hydroxide
⅓ oz (10 g) potassium carbonate

1 Weigh the coconut, castor, and olive oils and put them in a stainless steel saucepan. Weigh the water and pour into a plastic bucket. Half fill the larger saucepan with tap water and put it on the heat.

2 Put on your gloves, goggles, and mask. Weigh the potassium hydroxide and potassium carbonate separately and then combine. Add the potassium hydroxide and potassium carbonate mixture to the water in the bucket and stir for two minutes.

3 Melt the oils on a low heat and stir. Remove the soap pot from the heat. Still wearing your protective kit, pour the alkali mixture into the pot and stir again. Bring the soap to trace using a stick blender. (The oils and liquids may separate—keep mixing until they come together again.)

4 Place the soap pot into the larger pot and bring the water to a rolling boil. Cover the pot. Top off the water in the outer pot as it evaporates. Stir every 15 minutes for approximately three hours or until the mixture in the pot resembles deep yellow petroleum jelly.

5 Remove from the heat and test (see page 95). If the soap is ready, store in a plastic lidded container or dilute as per instructions on page 97. If not, continue to cook until it tests clear. Sterilize the foamer bottle.

Family Shampoo

In hard water areas, use diluted lemon juice or cider vinegar as a final rinse to restore balance and prevent the hair from clagging. Those with dry hair can use a natural conditioner after cleansing such as pure shea butter—rub into the roots, leave for ten minutes, and then rinse off.

1 Weigh the coconut, olive, castor, shea, and jojoba oils and put them in a stainless steel saucepan. Weigh the water and pour into a plastic bucket. Half fill your larger saucepan with tap water and put it on the heat.

2 Put on your gloves, goggles, and mask. Weigh the potassium hydroxide and potassium carbonate separately and then combine. Add the potassium hydroxide and potassium carbonate mixture to the water and stir for two minutes.

3 Melt the oils on a low heat and stir. Remove the soap pot from the heat. Still wearing your protective kit, pour the alkali mixture into the pot and stir again. Bring the soap to trace using a stick blender. (The oils and liquids may separate—keep mixing until they come together again.)

4 Place the soap pot into the larger pot and bring the water to a rolling boil. Cover the pot. Top off the water in the outer pot as it evaporates. Stir every 15 minutes for approximately three hours or until the mixture in the pot resembles deep yellow petroleum jelly.

5 Remove from the heat and test (see page 95). If the soap is ready, store in a plastic-lidded container or dilute as per the instructions below. If not, continue to cook until it tests clear.

INGREDIENTS FOR UNDILUTED SOAP PASTE

8 oz (200 g) coconut oil

8½ oz (210 g) olive oil

2 oz (50 g) castor oil

1 oz (20 g) shea oil

1 oz 20 g) jojoba oil

7 oz (180 g) water

4¼ oz (104 g) potassium hydroxide

⅓ oz (10 g) Potassium carbonate

1 drop food colorant per 10 oz (250 g) diluted soap

DILUTING, FRAGRANCING, AND THICKENING

For a 8 fl oz (250 ml) bottle of shampoo, sprinkle 1 tsp of cellulose acetate into 6 oz (166g) of boiling water and whisk. Add 3 oz (83 g) of soap paste and leave to dissolve, stirring from time to time. Heat the mixture to tepid, add a drop of coloring (if required), then add your chosen essential oil blends.

VARIATIONS

NORMAL HAIR	GREASY HAIR	DRY HAIR	FINE HAIR	ANTI-DANDRUFF
12 drops lemon	24 drops basil	36 drops lavender	48 drops chamomile	48 drops peppermint
24 drops geranium	24 drops lemon	24 drops rosemary	12 drops lavender	12 drops sage
24 drops parsley	12 drops thyme			

Shampoodle

I use this recipe as a dog wash for my two poodles. The essential oil blend
will help to deter fleas. You can moisturize a dog coat by rubbing
in some shea butter before the final rinse.

INGREDIENTS FOR UNDILUTED SOAP PASTE

8 oz (200 g) coconut oil

8 oz (200 g) canola
(rapeseed) oil

4 oz (100 g) olive oil

7 oz (180 g) spring water

4 oz (103 g) potassium hydroxide

⅓ oz (10 g) potassium carbonate

1 Weigh the coconut, canola, and olive oils and put them in a stainless
steel saucepan. Weigh the water and pour into a plastic bucket. Half fill
the larger saucepan with tap water and put it on the heat.

2 Put on your gloves, goggles, and mask. Weigh the potassium hydroxide
and potassium carbonate separately and then combine. Add the
potassium hydroxide and potassium carbonate mixture to the water
and stir for two minutes.

3 Melt the oils on a low heat and stir. Remove the soap pot from the heat.
Still wearing your protective kit, pour the alkali mixture into the pot and
stir again. Bring the soap to trace using a stick blender. (The oils and
liquids may separate—keep mixing until they come together again.)

4 Place the soap pot into the larger pot and bring the water to a rolling
boil. Cover the pot. Top off the water in the outer pot as it evaporates.
Stir every 15 minutes for approximately three hours or until the mixture
in the pot resembles deep yellow petroleum jelly.

5 Remove from the heat and test (see page 95). If the soap is ready,
store in a plastic lidded container or dilute as per the instructions
on page 97. If not, continue to cook until it tests clear.

VARIATIONS

BLEND ONE	BLEND TWO
36 drops cedarwood	24 drops citronella
24 drops lavender	24 drops thyme
	12 drops lavender

Rebatched Liquid Soap

If you don't want to make your liquid soap from scratch, you have two options: you can rebatch all your leftover bits of bar soap or you can dilute the soap paste you produce when you are making hot processed soap. Both methods are given below.

FROM SOAP SCRAPS

1 Grate the soap scraps and put in a jug. Boil 7 oz (200 g) of the water and pour it over the soap. Whisk until dissolved.

2 Add the remaining water and glycerine and mix with a stick blender until smooth. (Add more spring water at room temperature to make the soap thinner, if desired.) Allow to cool and then pour into a sterilized bottle.

FROM HOT PROCESS SOAP PASTE

1 When making bar soap using the hot process method (such as Laurus Nobilis on page 63), pour a couple of tablespoons of soap paste into a jug before pouring the rest into a mold.

2 Add three parts boiling water to one part paste and leave to dissolve. This makes a wonderful, thick, castile liquid soap.

INGREDIENTS

4 oz (100g) cold process soap scraps

12 oz (300g) distilled or spring water

1 tsp glycerine

Creams, Lotions, and Balms

THE BASICS

IF WE BELIEVE WHAT WE ARE TOLD BY the world of advertising, there is a cream available for absolutely everything. There are different creams, lotions, and balms for different parts of our body, for day and night, and many of these products come with a terrifying price tag. However, there are in fact only two types of cream—heavy creams, which are absorbed slowly to form a protective barrier (night creams, creams with a healing element, foot creams, and cleansers); and light creams, which absorb quickly (day moisturizers and lotions). Lotions are simply diluted creams, and balms are blends of fats, oils, and waxes that do not contain water.

All creams are a combination of botanical oils, essential oils, and water and are designed to penetrate and nourish the surface of the skin. Water and oils will not blend together, so a vegetable-based emulsifier is used to enable blending. Vegetable-based cetyl alcohol is also used to stabilize the mixture, and natural glycerine is added for its humectants properties.

If you enjoy making mayonnaise, you will love making creams and lotions, and by creating them yourself, you can be sure that you feed your skin with the natural goodness it deserves. Not only are storebought creams, balms, and lotions expensive, but many also contain petroleum-based products and other undesirable ingredients.

WHAT IS SKIN?

Skin is the largest organ in our bodies. An average-sized adult has about 2 square meters of skin weighing 6 lbs (2.7 kg), and 70 percent of our skin is made up of water. Our skin protects us from hostile environments, maintains our temperature, and helps our immune systems to combat disease. It also communicates sensation and excretes toxins that can affect our general health. In short, our skin is well worth taking care of. Structurally, the skin is made up of three principal layers and a large number of sub-layers. The principal layers are:

Epidermis: the outer layer of the skin (about the thickness of a sheet of paper)—this is the layer we see.

Dermis: this is the thick middle layer containing collagen and elastin which give skin strength and elasticity.

Subcutaneous: this is where fat, nerves, and blood vessels live; it is also home to the roots of hair follicles and oil and sweat glands.

WHAT BENEFITS DO CREAMS HAVE?

The basic ingredients used to make natural creams can be divided into three categories:

Humectants: attract water from the dermis and help to maintain moisture in the epidermis. Vegetable glycerine is the most commonly used natural humectant.

Occlusive agents: oils and waxes that moisturize by forming a physical barrier on the epidermis to act against water loss.

Emollients: lubricate and soften the skin by smoothing flaky cells. Natural emollients include shea butter and cocoa butter.

In addition to the above, essential oils and plant extracts, with their nourishing and antibacterial properties, are used. These are absorbed through the layers of our skin to help to sooth inflammation and encourage the growth of new skin cells.

TEXTURE

The texture of your creams and lotions, their feel on your skin, and their benefits are entirely dependent on the oils and active ingredients you select. The properties of specific oils and other active ingredients are listed on pages 10–14 to help you make a balanced selection for your recipes. Oils, fats, and butters containing a high proportion of saturated fatty acids (such as shea and cocoa butter) make heavy creams that sit on the surface of the skin. Less oily creams can be produced using oils such as hempseed, evening primrose, or sunflower oil, which are all absorbed easily.

PRESERVATIVES

Sadly, to date, an entirely natural preservative does not exist. Traditionally, parabens have been used to preserve face creams, but their use is controversial, and there are safer alternatives available on the market.

Whenever water is introduced into a cosmetic product, it is advisable to use a preservative to eliminate bacterial growth. If you intend to sell your creams, you need to guarantee a decent shelf life, so the use of a preservative is critical. If you are making creams for yourself without adding a preservative, make in small batches and store in the fridge.

When selecting a preservative, you need to ensure that it is fit for use with the product you intend to make. Some preservatives will not work when the end product has a high pH or where detergent is included in the formula. I recommend using optiphen (phenoxyethanol ethylhexylglycerin) to preserve products because it is effective and does not irritate sensitive skins. The recommended dose is 6 ml preservative per kilogram of product.

In addition to a preservative, you should always include an antioxidant such as vitamin E in your creams. While the preservative serves to keep the water content free of bacteria, the antioxidant reduces the potential of rancidity in the oils.

EMULSIFIERS

Water and oil will not mix, and emulsifiers are used to combine them into a stable emulsion. Also known as E-wax, most emulsifiers are vegetable-based and are a combination of polysorbate 20 (made from lauric acid/coconut oil), cetearyl alcohol, or ceteareth 20 (both plant-based fatty alcohols).

Most emulsifiers are sold as white, waxy pellets that you add to the oil stage of your formula. However, some E-waxes are divided into two parts, one of which you add to the water stage and one to the oil stage. Natural waxes such as beeswax and olive wax also act as emulsifiers, but they will thicken the product and are most commonly used in balms and ointments.

WATER

Use spring or distilled water in your creams and lotions—do not use tap water unless it has been put through a water filter.

ESSENTIAL OILS

Essential oils are added to creams both for their fragrance and their curative properties, and the amount added to a face cream is, on average, between 0.5 percent and 0.75 percent and should never exceed 1 percent. If you are using a regular bottle with a dropper and working with 2 oz (50 g) of cream, 0.5 percent is equivalent to 6 drops of essential oil and 0.75 percent equals 9

drops of essential oil . However, essential oils must be treated with the greatest respect, and the amount we use depends very much on the nature of the oil itself. Aromatherapy is a study of its own, so without some in-depth research, I would suggest that as a beginner you do not exceed 0.5 percent essential oils in your face cream formula. For fragrance, stick to plant hydrosols or hydrolats (used in place of your water) as an alternative. (See the Essential Oil Selection Chart on page III to get a an idea of which essential oils are best for which skin types.)

EQUIPMENT
Kitchen scales
While you can work with regular kitchen scales, you can also buy miniature scales that give you weights of less than a gram (fractions of an ounce). These are incredibly useful for cream makers.

Plastic jug
To weigh your ingredients in.
Medium-sized bowls x 2
Preferably stainless steel for mixing your formula in.
Saucepans x 2
You want to create two bain-marie set ups, so ideally you should be able to sit your bowls comfortably on the rim of the saucepans.
Aluminium foil
To cover the bowls while heating.

Thermometer
A digital prober is ideal.
Whisk
An electric whisk will make your life easier.
Measuring spoons
Plastic spatula
Paper towels
Washing-up bowl
Vinegar
For sterilizing equipment.
Heat source
Plastic pipettes
For essential oils.
A sponge
For cleaning up.
Assorted containers
For your finished products.

HYGIENE
When making creams, cleanliness and sterility are two major issues. While it is perfectly possible to make creams in your kitchen, you should adopt a regime to ensure that bacteria does not enter your products.

Ensure your hands are clean and your clothes are protected. It's a good idea to not wear woolly jumpers when making creams since these can shed into your product. Tie back your hair or cover with a net or scarf.

Fill a washing up bowl with boiling water for sterilizing, and add two tablespoons of vinegar. Soak all your utensils and jars in this solution for five minutes, rinse thoroughly, and then leave them to dry on paper towels (jars bottoms up). Immerse a new sponge in the sterilising bowl and wipe down your work surfaces, then cover with paper towels. Always keep all the equipment you use separate from your day to day cooking equipment.

The biggest cause of bacteria in creams is finger dipping! When applying creams, take them out of the jar with a small spatula. Store your cream making oils in the fridge and keep your essential oils in dark bottles or in a dark cupboard.

MAKING CREAMS & LOTIONS

Creams are made in four distinct stages:

Stage 1: the oil stage

Melt the oils, waxes, one-stage emulsifiers, and cetyl alcohol in a heat-resistant jug or stainless steel bowl on a double boiler. Mix until smooth and heat to around 167°F (75°C).

Stage 2: the water stage

Put the water, glycerine, preservative (optional) and/or second-stage emulsifier (if you are using one) into another jug or bowl, mix until smooth, then heat to 167°F (75°C) on a second double boiler. Take both saucepans off the heat and pour the oil into the water in a slow stream while whisking.

Stage 3: active additives

Once the temperature of your mixture has dropped below 104°F (40°C), add the vitamin E oil, plant extracts, or other active ingredients to your cream.

Stage 4: essential oils

Wait until the temperature drops below 86°F (30°C), then add the essential oils. Add a drop of food or cosmetic colorant, if required.

Pour (or use a sterilized spatula to scrape your cream/lotion) into sterilized pots/bottles. Cover with a paper towel until cold and then secure the lids. Do not put the lid on when the formula is still warm since bacteria-attracting condensation can get trapped inside.

CREATING RECIPES

The chart on page 110 gives guidelines for selecting oils. The percentages of ingredients used in your creams should be determined by the desired consistency. To create a rich, heavy cream suitable for dry skin, the total content of fats/oils in your formula should be in the region of 35 percent. For a general cream, select ingredients for dry skin on the chart but bear in mind that both cocoa butter and shea butter will increase the cream's thickness. For a heavy, mature skin cream, select 35 percent of your formula from the oils indicated for mature skin. The ingredients listed for oily skin will give you a lighter textured cream. If you have dry or mature skin but would like a light cream, avoid the inclusion of shea or cocoa butter or reduce the fat and increase the water content until you reach your ideal texture. As an alternative, you can replace a percentage of the water content with aloe vera gel, flower waters, or witch hazel. Lotions can be made successfully by simply beating up to 50 percent cold spring water into your completed cream. If you want to color your cream, add a single drop of food or cosmetic colorant during the final stage.

CALCULATING FORMULATIONS

Recipes for creams, balms, and lotions should always be calculated as percentages, so when you add up the percentages for each ingredient in your formula, you should have a total of 100 percent.

As you are often weighing very small amounts of ingredients, some cream makers spend a lot of time with tables that convert mililiters to grams or vice versa, but I find it simpler to formulate all cream ingredients in grams to give me 4 oz (100 g) in total, and then to halve or multiply the formula to provide smaller or larger amounts. Sometimes with very small amounts you might find yourself a fraction over or under the 100 percent, but I have not found this to be an issue. As a rule of thumb, creams and lotions contain:

- 20%–35% oils and fats
- 60%–75% water, gels, or hydrosols
- 6%–8% emulsifiers
- 0.5%–1% essential oils

The following are useful guides for ingredients that are difficult to weigh:

- 12–15 drops of essential oil (these vary from oil to oil but the viscosity should give you an idea) = 0.5 g
- 1 ml vitamin E (use a pipette for this) = 1 g
- 12 x drops liquid preservative = 0.5–0.6 g

BASE INGREDIENTS	OILY	DRY	COMBINED	PROBLEM	MATURE	HIGH SAT ACIDS
Almond (sweet)		✓	✓	✓	✓	
Aloe Vera	✓	✓	✓	✓	✓	
Apricot Kernel	✓	✓	✓	✓	✓	
Argan		✓		✓	✓	
Avocado	✓	✓	✓	✓	✓	✓
Baobab		✓			✓	
Borage				✓	✓	
Black cumin				✓	✓	
Coconut (virgin)	✓		✓			✓
Calendula				✓		
Carrot Seed	✓	✓	✓	✓	✓	
Castor Oil	✓			✓		
Cocoa Butter		✓	✓	✓	✓	✓
Comfrey				✓		
Evening Primrose		✓	✓	✓	✓	
Glycerin (veg)		✓		✓		
Grapeseed	✓		✓	✓		
Hempseed	✓		✓	✓	✓	
Jojoba	✓	✓	✓	✓	✓	
Macadamia					✓	
Melon	✓	✓	✓	✓		
Meadowfoam	✓				✓	
Olive		✓		✓	✓	
Peach Kernel	✓	✓	✓	✓	✓	
Rosehip Seed		✓		✓	✓	
Sea Buckthorn			✓		✓	
Sunflower Oil	✓	✓	✓	✓	✓	
Shea Butter		✓	✓	✓	✓	✓
Wheatgerm Oil		✓		✓	✓	
Witch Hazel	✓	✓	✓	✓	✓	
Vitamin E		✓		✓	✓	

SKIN CONDITIONS

On both charts on pages 110–111, I have highlighed oils that are proven to be gentle and, in some cases, helpful to problem skins. Many people, however, look to natural cosmetics for a cure to serious conditions, such as severe eczema or psoriasis, for which there is no known cure. If you suffer from mild forms of these conditions, it is likely that your skin will benefit from products that are free from harmful chemicals, but for serious conditions such as dermatitis, it is recommended that you seek medical advice.

If you are considering making products for babies, use oils such as borage or calendula, and while a chamomile hydrolat could be added in small quantity, avoid the use of essential oils altogether.

ESSENTIAL OILS SELECTION CHART BY SKIN TYPE

ESSENTIAL OILS & HYDROSOLS	OILY	DRY	SENSITIVE	PROBLEM	MATURE	AVOID DURING PREGNANCY
Chamomile (Roman)		√	√			
Cypress		√				Avoid months 1–5
Eucalyptus	√			√	√	
Geranium	√	√			√	Avoid months 1–3
Grapefruit	√			√		
Helichrysum		√	√	√	√	
Jasmine	√		√			√
Lavender	√	√		√		√
Lemon	√	√			√	
Mandarin	√			√		
Neroli			√			
Palmarosa		√				
Patchouli		√			√	
Peppermint	√			√		√
Rose Absolute		√			√	√
Rose Geranium				√		
Rosemary	√			√		
Sandalwood		√				
Teatree	√	√		√		
Ylang-ylang	√	√			√	

BASIC CREAMS AND LOTIONS METHOD

1. Weigh and stir the ingredients for the oil phase in a bowl or heat-proof jug. Weigh and stir the ingredients for the water phase in a separate bowl or jug. Put both on the heat and heat to 167°F (75°C).

2. Pour the oil phase into the water phase in a steady stream, whisking continuously.

3. Once the mixture is cool and thick, add the essential oils, whisk again, and scoop into sterilized pots.

CREAMS

ONE OF THE TRUE DELIGHTS OF making creams, lotions, and balms is the absence of caustic soda. You can use the best quality oils with the certain knowledge that their therapeutic properties will remain active.

The following recipes will produce around 3½ oz (100 g) of product, but you are free to halve or double quantities as you wish. The important thing is that the total of all the ingredients adds up to as close as you can get to 3½ oz (100 g). Use the charts on pages 110–111 to substitute oils according to the particular properties they will bring to your product.

You can also substitute spring water for flower waters, herbal infusions, teas, or aloe vera, which will add a fluffy texture to your cream. It is a simple balancing act—if you want to add something that isn't in the recipe, be sure to reduce the quantity of another oil or liquid in the recipe by the same amount, and remember that a higher water content means a thinner cream or lotion.

Sleeping Beauty

HEAVY CREAM

This delicious night cream will nourish and soften so you wake up with supple and silky skin. It's particularly good for dry skin and is fragranced with lavender essential oil, which should help you sleep well, but use geranium instead if you are at all stressed out.

1 Sterilize all your equipment and pots. Weigh the ingredients for the oil phase, put them in a bowl, and mix. Weigh the ingredients for the water phase, put them in another bowl, and mix.

2 Put both bowls on separate bain-maries (double boilers), covering the bowls with silver foil. Heat both to 167°F (75°C).

3 Remove from heat and steadily pour the oil phase into the water phase in a continuous stream, whisking constantly. When the temperature reduces to 95°F (35°C) stir in the vitamin E oil. When the temperature reaches 86°F (30°C) add the essential oil blend of your choice and stir.

4 Pour into prepared jars and cover with kitchen paper. Apply lids when cool.

INGREDIENTS

OIL PHASE

½ oz (12 g) jojoba oil

¼ oz (6 g) unrefined cocoa butter

¼ oz (5 g) avocado oil

⅛ oz (4 g) apricot kernel oil

⅛ oz (2 g) beeswax

⅛ oz (2 g) cetyl alcohol

⅛ oz (4 g) E wax

WATER PHASE

2 oz (50 g) spring water

⅛ oz (2 g) glycerine

½ oz (11 g) aloe vera gel

1/16 oz (1 g) preservative

ADDITIVES

1/16 oz (1 g) vitamin E oil

ESSENTIAL OILS

24 drops lavender

VARIATION

Replace the lavender oil with the same quantity of geranium oil for a soothing scent.

Fragrant Feet

HEAVY CREAM

This cream is a deep rub that is designed to soften rough skin and deodorize and restore your feet at the end of the day. If you have cracked heels, rub generous amounts of the cream into your feet just before you go to bed and sleep with socks on.

INGREDIENTS

OIL PHASE

½ oz (10 g) olive oil
⅓ oz (8 g) shea butter
¼ oz (5 g) virgin coconut oil
¼ oz (5 g) sunflower oil
⅛ oz (2 g) cetyl alcohol
¼ oz (5 g) E wax

WATER PHASE

2½ oz (61 g) spring water
⅛ oz (2 g) glycerine
1/16 oz (1 g) preservative

ADDITIVES

1/16 oz (1 g) vitamin E oil
1 drop colorant (optional)

ESSENTIAL OILS

4 drops tea tree
2 drops rosemary

1. Sterilize all your equipment and pots. Weigh the ingredients for the oil phase, put them in a bowl, and mix. Weigh the ingredients for the water phase, put them in another bowl, and mix.

2. Put both bowls on separate bain-maries (double boilers), covering the bowls with silver foil. Heat both to 167°F (75°C).

3. Remove from heat and steadily pour the oil phase into the water phase in a continuous stream, whisking constantly. When the temperature reduces to 95°F (35°C), stir in the vitamin E oil. When the temperature reaches 86°F (30°C), add the essential oil blend of your choice and colorant, if using, and stir.

4. Pour into prepared jars and cover with kitchen paper. Apply lids when cool.

VARIATION

Replace the essential oils with 6 drops of cypress and 6 drops of lavender and add a tablespoon of pumice powder or oatmeal to your cream.

Baby's Bottom

HEAVY CREAM

This rich cream contains a selection of oils that will be kind on a baby's bottom. It's a good cream to use to prevent diaper rash because it will form a light barrier to protect the skin. I've used one drop of blue and red food coloring for each jar.

1 Sterilize all your equipment and pots. Weigh the ingredients for the oil phase, put them in a bowl, and mix. Weigh the ingredients for the water phase, put them in another bowl, and mix.

2 Put both bowls on separate bain-maries (double boilers), covering the bowls with silver foil. Heat both to 167°F (75°C).

3 Remove from heat and steadily pour the oil phase into the water phase in a continuous stream, whisking constantly. When the temperature reduces to 95°F (35°C), stir in the vitamin E oil. When the temperature reaches 86°F (30°C), add the borage oil and stir.

4 Pour into prepared jars and cover with kitchen paper. Apply lids when cool.

INGREDIENTS

OIL PHASE
½ oz (10 g) olive oil
½ oz (12 g) calendula oil
½ oz (10 g) shea butter
⅛ oz (2 g) cetyl alcohol
⅛ oz (4 g) E wax

WATER PHASE
2¼ oz (54 g) chamomile water
⅛ oz (2 g) glycerine
1/16 oz (1 g) preservative

ADDITIVES
1/16 oz (1 g) vitamin E oil
⅛ oz (4 g) borage oil

VARIATION
If baby already has a diaper rash, replace the borage oil with comfrey oil and leave out the colorant.

Almond and Calendula Hand Cream

LIGHT CREAM

To make calendula oil, simply fill a half pint (half liter) jar or bottle one-third with calendula petals, top off with olive oil, and leave it on a sunny windowsill for a week, shaking from time to time. At the end of the week, sieve out the petals, and the oil is ready to use.

INGREDIENTS

OIL PHASE
⅓ oz (8 g) almond oil
⅓ oz (8 g) calendula oil
¼ oz (7 g) hempseed oil
⅛ oz (2 g)wheat germ oil
⅛ oz (2 g) cetyl alcohol
⅓ oz (8 g) E wax

WATER PHASE
2½ oz (61 g) spring water
⅛ oz (2 g) glycerine
1⁄16 oz (1 g) preservative

ADDITIVES
1⁄16 oz (1 g) vitamin E oil

ESSENTIAL OILS
6 drops lavender
4 drops ylang-ylang
2 drops rose

1 Sterilize all your equipment and pots. Weigh the ingredients for the oil phase, put them in a bowl, and mix. Weigh the ingredients for the water phase, put them in another bowl, and mix.

2 Put both bowls on separate bain-maries (double boilers), covering the bowls with silver foil. Heat both to 167°F (75°C).

3 Remove from heat and steadily pour the oil phase into the water phase in a continuous stream, whisking constantly. When the temperature reduces to 95°F (35°C), stir in the vitamin E oil. When the temperature reaches 86°F (30°C), add the essential oil blend, and stir.

4 Pour into prepared jars and cover with kitchen paper. Apply lids when cool.

VARIATION
Try replacing the essential oils with 6 drops of patchouli and 6 drops of geranium.

Aloe Vera Cream

LIGHT CREAM

Gorgeous, cooling aloe vera is a fine additive for your creams and lotions and creates a fluffy texture. Use as a replacement for some of the water content. You can replace the watermelon and baobab oils with almond and jojoba oil, if you prefer.

1 Sterilize all your equipment and pots. Weigh the ingredients for the oil phase, put them in a bowl, and mix. Weigh the ingredients for the water phase, put them in another bowl, and mix.

2 Put both bowls on separate bain-maries (double boilers), covering the bowls with silver foil. Heat both to 167°F (75°C).

3 Remove from heat and steadily pour the oil phase into the water phase in a continuous stream, whisking constantly. When the temperature reduces to 95°F (35°C), stir in the vitamin E oil. When the temperature reaches 86°F (30°C), add the essential oil blend of your choice and stir.

4 Pour into prepared jars and cover with kitchen paper. Apply lids when cool.

INGREDIENTS

OIL PHASE
½ oz (12 g) watermelon oil
¼ oz (7 g) baobab oil
⅛ oz (2 g) cetyl alcohol
¼ oz (5 g) E wax

WATER PHASE
1½ oz (40 g) spring water
⅛ oz (2 g) glycerine
1 oz (22 g) aloe vera gel
1/16 oz (1 g) preservative

ADDITIVES
1/16 oz (1 g) vitamin E oil
⅓ oz (8 g) evening primrose oil

ESSENTIAL OILS
12 drops essential oils
of your choice

VARIATION

For a refreshing blend, use 6 drops each of rosemary and lemon essentail oils, or for problem skin try 4 drops each of mandarin, grapefruit, and rose geranium.

Sunflower and Argan

LIGHT CREAM

In France, argan oil is highly prized for its ability to rejuvenate tired skin. Combined here with sunflower and almond oils, this cream has a light texture, is easily absorbed, and will keep mature skin soft and supple.

INGREDIENTS

OIL PHASE
¾ oz (20 g) sunflower oil
⅓ oz (8 g) almond oil
¼ oz (6 g) argan oil
⅛ oz (2 g) cetyl alcohol
¼ oz (6 g) E wax

WATER PHASE
2¼ 55 g rose water
1⁄16 oz (1 g) glycerine
1⁄16 oz (1 g) preservative

ADDITIVES
1⁄16 oz (1 g) vitamin E oil

ESSENTIAL OILS
12 drops essential oils of your choice

1 Sterilize all your equipment and pots. Weigh the ingredients for the oil phase, put them in a bowl, and mix. Weigh the ingredients for the water phase, put them in another bowl, and mix.

2 Put both bowls on separate bain-maries (double boilers), covering the bowls with silver foil. Heat both to 167°F (75°C).

3 Remove from heat and steadily pour the oil phase into the water phase in a continuous stream, whisking constantly. When the temperature reduces to 95°F (35°C), stir in the vitamin E oil. When the temperature reaches 86°F (30°C), add the essential oil blend of your choice and stir.

4 Pour into prepared jars and cover with kitchen paper. Apply lids when cool.

VARIATION
Try an essential oil blend of 6 drops helichrysum, 4 drops neroli, and 2 drops rose or 6 drops geranium, 4 drops lavender, and 2 drops lemon.

Flower Milk

LIGHT CREAM

There is something extraordinarily satisfying about collecting flowers from your garden, creating flower water, and then whipping it up into a cream. Flower milks are great for daily skin cleansing and can be followed by a splash of neat flower water.

INGREDIENTS

OIL PHASE

¼ oz (6 g) almond oil

¼ oz (6 g) virgin coconut oil

⅛ oz (2 g) apricot kernel oil

⅛ oz (2 g) cetyl alcohol

⅛ oz (4 g) E wax

WATER PHASE

3 oz (75 g) flower water (see opposite)

⅛ oz (3 g) glycerine

¹⁄₁₆ oz (1 g) preservative

ADDITIVES

¹⁄₁₆ oz (1 g) vitamin E oil

ESSENTIAL OILS

12 drops essential oils of your choice

2 drops of colorant of your choice

1 Sterilize all your equipment and pots. Weigh the ingredients for the oil phase, put them in a bowl, and mix. Weigh the ingredients for the water phase, put them in another bowl, and mix.

2 Put both bowls on separate bain-maries (double boilers), covering the bowls with silver foil. Heat both to 167°F (75°C).

3 Remove from heat and steadily pour the oil phase into the water phase in a continuous stream, whisking constantly. When the temperature reduces to 95°F (35°C), stir in the vitamin E oil. When the temperature reaches 86°F (30°C), add the essential oil blend of your choice and stir. If you prefer a thinner lotion, dilute with boiled, cooled spring water until the desired consistency is reached.

4 Pour into prepared bottles and cover with kitchen paper. Apply lids when cool.

VARIATION

To make lavender milk, use lavender water and essential oil, or substitute both the water and oils for rose or chamomile.

Flower Water

Steam distillation can produce the most wonderful flower waters that can be added to creams and lotions. Lavender, fresh bay leaves, and rose petals all make lovely flower waters, which can be made easily using a few pots and bowls on your kitchen stove.

1 Put the stockpot on the stove (do not turn on the heat source yet). Place the "stand" inside the stockpot. Fill the area around and to the level of the stand with petals or herbs. Pour spring water over the petals/herbs to the level of the stand.

2 Place your heatproof bowl right side up on top of the stand. Turn the heat source to high. Put the stockpot lid upside down on top of the stockpot. Fill the inverted stockpot lid with ice.

3 As the water heats up, the steam will rise. As it hits the ice, the steam will drip from the upside down saucepan lid into the empty bowl inside to make flower water. When the ice melts, empty the saucepan lid and refill it with more ice. Continue the process until all the water at the bottom of the pan has disappeared—take care to not let the pan burn!

4 If you want to preserve the flower water, you will need to add an equal quantity of alcohol (such as vodka) or denatured alcohol. Personally I do not like using alcohol in my creams because it can make them curdle. I prefer to leave my flower water virgin and add a branded preservative during the cream-making process.

INGREDIENTS

spring water

petals or herbs (fresh or dried)

jug of ice cubes

EQUIPMENT

tall stainless steel stockpot with (preferably glass) lid

heatproof glass or stainless steel bowl that will fit inside the stockpot

something to stand the bowl on in the pot (you could use another small inverted bowl or even a brick)

NOTE

I haven't included sizes or quantities for the ingredients because it depends how much flower water you want to make. The important thing is that the pot needs to be big enough to hold the bowl on its "stand" and to allow the pot lid to be placed upside down on the top without leaving any gaps.

BALMS

A BALM IS A GENTLE NONAQUEOUS cream designed to both protect and heal the skin. It is made simply by melting together a combination of natural oils, fats, and waxes in a double boiler (bain-marie) and adding essential oils. Each ingredient is selected for its specific therapeutic property.

In addition to the wide wariety of vegetable oils available, try experimenting with a range of waxes and butters. The bee has been generous enough to supply us with honey, beeswax, propolis, and royal jelly—all of which could be included in a balm. (Vegans may prefer to use olive wax, Mexican candelilla wax, and/or rice bran wax flakes.) You can also include exotic butters in your blends.

Top of the list are cocoa and shea butter, but if you can obtain mango, avocado, muru muru, capuçau (from Brazil), or even coffee butter, try these, especially in lip balms. These recipes will make 3½ oz (100 g) of product, so if that is what you want to achieve, keep the basic numbers the same.

Since balms do not contain water, you do not need to use a preservative, but you should add 1 percent vitamin E oil as an antioxidant that will protect the oils in the formula from rancidity.

All-Purpose Balm

This is a basic recipe that will help to soften and protect the skin. You can substitute the beeswax with candelilla or olive wax, and you can substitute the oils using the chart on page 110 for guidance. You can also infuse your oils with herbs.

1 Sterilize the jars. Weigh all the ingredients. Using a saucepan on a double boiler, heat all the stage 1 ingredients to melting point.

2 Remove the saucepan from the heat and whisk the mixture until it cools to 104°F (40°C).

3 Add the stage 2 ingredients. Mix well. Pour into jars and cover with paper towels. Apply lids when cool.

INGREDIENTS

STAGE 1

½ oz (14 g) beeswax

1½ oz (40 g) olive oil

½ oz (18 g) jojoba oil

¾ oz (21 g) calendula oil

¼ oz (6 g) shea butter

STAGE 2

12 drops vitamin E oil

20 drops essential oil
(see Variations below)

VARIATIONS

BRUISING	DEEP HEAT RUB	INSECT BITES
Replace the calendula oil with comfrey oil	4 drops lavender	6 drops Roman chamomile
3 drops lavender	4 drops ginger	6 drops lavender
3 drops rosemary	3 drops eucalyptus	
3 drops geranium	1 drop clove bud	
3 drops arnica		

Gardeners' Balm

This rich antiseptic balm will protect the hands while gardening and soothe any scratches. It is also useful as a general antiseptic balm. The watermelon and baobab oils can be substituted with almond and jojoba, and the muru muru butter with another butter of your choice.

INGREDIENTS

STAGE 1

½ oz (14 g) beeswax

1½ oz (40 g) olive oil

1 oz (24 g) watermelon oil

1 Tbsp (15 g) baobab oil

¼ oz (6 g) muru muru butter

STAGE 2

12 drops vitamin E oil

10 drops tea tree oil

5 drops rosemary oil

5 drops eucalyptus oil

1 Sterilize the jars. Weigh all the ingredients. Using a saucepan on a double boiler, heat all the stage 1 ingredients to melting point.

2 Remove the saucepan from the heat and whisk the mixture until it cools to 104°F (40°C).

3 Add the stage 2 ingredients. Mix well. Pour into jars and cover with paper towels. Apply lids when cool.

VARIATION

To keep the bugs at bay while you are working, replace the essential oil blend with 5 drops each of lavender, thyme, eucalyptus, and rosemary.

Lip Balms

When making lip balms, texture and taste are important. We want to achieve a balm that will soften but also protect the lips, and for this reason, I like to work with chocolatey, unrefined cocoa butter and delicious olive wax.

1 Sterilize the jars. Weigh all the ingredients. Using a saucepan on a double boiler, heat all the stage 1 ingredients to melting point.

2 Remove the saucepan from the heat and whisk the mixture until it cools to 104°F (40°C).

3 Add the vitamin E oil and the essential oil blend. Mix well. Pour into jars and cover with paper towels. Apply lids when cool.

INGREDIENTS

STAGE 1

1 oz (28 g) olive wax

¾ oz (21 g) calendula oil

¾ oz (20 g) cocoa butter

½ oz (16 g) castor oil

½ oz (14 g) apricot kernel oil

STAGE 2

12 drops vitamin E oil

20 drops essential or fragrance oil of your choice

VARIATIONS

CHOCOLATE VANILLA

Add 1 tsp cocoa powder to the first stage and 4 drops vanilla extract to the second stage for a lip lickable experience.

HONEY BEE

Add ½ tsp honey during stage 1 and use a honey fragrance oil.

MANGO TANGO

Replace the cocoa butter with mango butter and use a mango fragrance oil or geranium essential oil.

WHIPS AND MELTS

THE FOLLOWING RECIPES GIVE YOU the opportunity to seriously indulge yourself and ensure that your skin remains soft and lovely. Run a steaming hot bath, light a scented candle, and treat your skin with some of these fantastic products.

The Shea Butter Whip is designed to be slathered on your whole body after a warm bath. It will be quite solid to start with but will melt into your skin. Store in a wide-necked jar so you can scoop out handfuls.

Bath Melters are little chunks of moisturizing oils and butters that will make the bathwater super soft, and salts from the Dead Sea are packed with minerals. Try adding them to your bathwater or mix them with oil into a skin scrub.

Bath oils offer the best way of adding fragrance to your bath and are extremely simple to make. For the ultimate indulgence, try rubbing yourself all over with a sugar or salt scrub to leave your skin exfoliated and polished.

Shea Butter Whip

This mouthwatering blend of oils and butters is designed to melt on warm (after bath) skin. The shea butter can be substituted with a more exotic butter such as mango or muru muru, and you can also use this as a hair conditioner. This recipe will make 10½ oz (300 g) of whip.

1 Sterilize the jars. Weigh all the ingredients. Put the shea butter in a jug or bowl on a double boiler and heat just to melting point.

2 Remove from heat and add the coconut, melon, avocado, and vitamin E oils. Mix well. Add your chosen essential oil blend.

3 Stand the jug in a bowl of iced water and whip with either a hand or electric whisk until it has a creamy consistency (you can speed this up by putting it in the fridge for ten minutes before whipping).

4 Scoop into sterilized jars.

NOTE: Shea butter can crystallize due to temperature variations, and as a result your finished products may be grainy. To prevent this, heat your shea butter in a double boiler to 82°C (180°F) for 20 minutes before use, then put it in the fridge to solidify again and use as normal.

INGREDIENTS

8 oz (200 g) shea butter

2 oz (50 g) virgin coconut oil

1 oz (25 g) watermelon oil

1 oz (24 g) avocado oil

24 drops (1 g) vitamin E oil

20 drops essential oils (see Variations below)

VARIATIONS

NIGHT OUT
6 drops geranium
6 drops patchouli
6 drops mandarin
2 drops rose

MORNING GLORY
10 drops lime
3 drops sweet orange
3 drops Roman chamomile
2 drops neroli
2 drops jasmine

SLEEP WELL
12 drops lavender
8 drops Roman chamomile

Bath Melters

These little parcels of heaven will dissolve in your bath to soften and fragrance the water.
Use latex ice cube trays as molds as they give you just about the right size of melt.
This recipe makes 5 oz (150 g) of product (approximately 24 bath melters).

INGREDIENTS

4 oz (100 g) cocoa butter

¾ oz (20 g) almond oil

¼ oz (10 g) calendula oil

¼ oz (10 g) boabab oil

¼ oz (10 g) apricot kernel oil

20 drops essential oils (see Variations below)

1 Weigh your ingredients. Put the cocoa butter in a jug or bowl and bring to melting point on a double boiler.

2 Remove from heat and stir in the remaining oils. Add your chosen essential oil blend.

3 Pour into an ice cube tray and put in the fridge to harden.

VARIATIONS

FLIRTY

10 drops ylang-ylang

5 drops lavender

5 drops sweet orange

SNUFFLY

13 drops rosemary

5 drops eucalyptus

2 drops ginger

Salt and Sugar Scrubs

Before investing in an expensive spa treatment, bear in mind that you can buy the ingredients for salt and sugar scrubs cheaply and that the recipe is really simple—two parts salt or sugar (or a combination of both) to one part oil. Place in a wide-necked jar, add a few drops of your favorite essential oil and shake. That's it! Both the salt and sugar will sink to the bottom of the jar, so remember to shake before use.

The subtlety is in which type of salt or sugar you buy. It's not a good idea to go for large, sharp grains because these could be scratchy on the skin, so choose fine or medium grains. While you can make this product using regular table salt, if you want something that will be great for your skin, try using a combination of sea salt, Dead Sea salt, and Epsom salts, and the proportion you choose for each really doesn't matter. For sugar scrubs, I always use demerara, a medium-grained, honey-colored sugar. A teaspoon of honey added to the mix works well. When it comes to selecting oils, you could use simple sunflower oil or create a blend yourself using the oil chart on page 110. Jojoba and almond oil are good basic choices.

If you want the scrub to be cleansing as well as polishing, replace a third of the oil with some of your homemade liquid soap, or add a bit of storebought liquid castile soap. You could also throw in some dried herbs, both for fragrance and for additional exfoliation.

To use your scrub, scoop a handful of the mixture onto a bath mitt or washcloth and rub well into the skin in a circular motion, rubbing extra hard on rough areas such as heels and elbows. Rinse off thoroughly, and you should feel shiny, tingly, and brand new again.

Bath and Body Oil

A good, long soak in a bath containing bath oil will undoubtedly help to soften, nourish, and moisturize your skin since the heat of the bath helps the body to absorb the oils. You can use virtually any oil or combination of plant oils as a base or you can infuse an oil with dried herbs and benefit from their specific properties. The blend of essential oils you add to your base is the element that makes the bath oil special and also dictates the mood—if you want to relax, refresh, or indulge, select your essential oils accordingly.

A little essential oil goes a long way, so include the blend at just 1 percent the weight of your plant oils, or less. You can also use this as a massage oil.

INGREDIENTS

The following is a well-balanced base formula:

1 part sweet almond oil

1 part jojoba oil

1 part watermelon oil

NOTE: The following essential oil blends are for 3½ fl oz/ 100 ml base oil. Adjust quantities accordngly if you are making more or less product.

RELAX

8 drops chamomile

8 drops rose geranium

8 drops lavender

REFRESH

8 drops lemon

8 drops mandarin

8 drops grapefruit

INDULGE

8 drops patchouli

8 drops mandarin

8 drops ylang-ylang

SETTING UP A SOAPMAKING BUSINESS

SINCE I SOLD MY SOAP BUSINESS, I HAVE dedicated a lot of time to teaching people both the legal obligations and the business and personal skills they will need to make a soapmaking enterprise successful. Many people who come to me are hobbyists who are addicted to their craft and who feel that running a business is the next natural step; others see a soapmaking business as a way to work from home. By the end of the course, quite a number of the participants decide that the process is far more complicated and requires more time and commitment than they had anticipated, and instead continue making soap as a hobby. Others take the bit between their teeth and go on to build successful businesses. Your ability to run a successful soapmaking business lies entirely with you—your circumstances, your skill set, and, like any other business, your ability to plan ahead, understand the numbers, and most of all to build and market a successful brand. Can you make money from a handmade soap business? Yes, but only if you are the right person to run it.

While many believe that working for themselves will be an easy option, it is far from it. Building any business requires a huge amount of commitment, risk, and dedication, and unless you thrive on administration, this can be a time consuming and tedious task. Soapmaking is no different. While it is undoubtedly a rewarding, addictive, and creative hobby, you could find that the time and energy you need to spend building your business means a great deal less time can be spent actually doing what you love: making the soap. To run a successful soap business, you need to understand all the skills and tasks required, and if you don't have them or can't do them, make sure you know someone who can. Business is about the bottom line, planning is crucial and the ability to cost accurately and keep proper financial records is critical.

You need to adopt the mind-set that your soapmaking is no longer a hobby. Plan your time carefully and be warned: it is easy for a small business to entirely take over your life—but at least you have all those wonderful bath oils and soaks to relax with.

Soap is made with a caustic chemical that can harm you if used in excess in your products. In addition, some botanical ingredients and additives are not recommended for use on the skin. For these reasons, it is essential that you ensure that your business is selling safe and desirable products. Different countries have specific laws to ensure you take care of these responsibilities.

EUROPE Within the European Union, soap falls under Cosmetic Legislation. Full details can be found at http://ec.europa.eu/consumers/sectors/cosmetics/documents/guidelines/index_en.htm

USA Providing your soap does not claim to beautify or moisturize the skin, it will be classified as a detergent. However, if any claims at all are made, it falls under the laws of the Food and Drug Administration (FDA) and full details can be found at http://www.fda.gov/Cosmetics/GuidanceComplianceRegulatoryInformation/ucm074201.htm

CANADA Soap is classified as a cosmetic. You will find details of the legislation at http://www.hc-sc.gc.ca/cps-spc/person/cosmet/index-eng.php

AUSTRALIA Soap is classified as a cosmetic, and great emphasis is placed on labeling and the use of inaccurate claims. You will find full details at http://www.tga.gov.au/docs/html/cosclaim.htm

While complying with the legislation is time consuming, it is designed to ensure that your product is safe and that you can demonstrate complete traceability of its ingredients. You must allot a batch number to each batch of product, and you should also keep records of the suppliers and batch numbers of the ingredients you have used.

In practice, this means that if you are ever challenged by an individual who believes your product has given him/her a rash, or worse, you have enough information to be able to establish whether this could be the case. Be warned; we live in a litigious society, and compliance to the legislation is a good thing for everyone. Compliance also demonstrates to your customers that you are a responsible manufacturer who takes his/her craft seriously. While individual countries (within and outside of Europe) have their own additional trading laws that you must comply with, most countries accept that EU legislation is stringent and the standards are respected throughout the world.

The legislation states that a "responsible person" must assess and certify any soap or cream formulation that you make for sale. This means you cannot legally sell a single bar of soap—be it to a friend, through a shop, on the Internet, or via a school festival—unless the recipe is certified. The responsible person needs to be a qualified (and fully insured) cosmetic chemist, toxicologist, pharmacist, or GP. I have given a recommendation for a highly competent assessor on page 143.

The above rule applies to all bath products, including melt and pour soap, creams, lotions, and bath bombs. Your safety assessor will require the following information from you:

- A list of the ingredients with their International Nomenclature of Cosmetic Ingredients (INCI) names. This is usually the Latin name of the ingredient.
- The percentages of each ingredient within your recipe, starting with the largest.
- Material Safety Data Sheets (MSDS sheets) for all your ingredients (ask your supplier for these).
- Colorants with their CI numbers.
- Proof from your supplier that any fragrance oils or additives you use comply with International Fragrance Association (IFRA) guidelines.
- A sample of your label (see Labeling, below).
- Knowledge about your step-by-step manufacturing procedures.
- A written step-by-step description of your manufacturing procedures and of your workshop.

Some countries in Europe require that your workshop is completely separate to your living accommodation. This could be a room in your house dedicated to your soapmaking.

Cosmetic legislation also requires you to keep a comprehensive Product Information File (PIF) for each of your formulations. This should contain your recipe sheet with its allocated name, batch number, and date of manufacture. Write on your recipe sheet the supplier of the ingredient and their batch number for it, the size and shape of the finished product, description of any tests you undertook to ensure your product is stable, pH reading if applicable, how you packaged the product, and notes on any issues that occurred during manufacture. You should also keep records of who you sold the batch to if at all possible.

LABELING

The law states that consumers must be given access to information on all of the ingredients within the product at point of sale. If you are wrapping your product, then these details must be on the label. If not, they can be stated on a card near the product. The ingredients should be headed "INGREDIENTS" in caps and listed in INCI format from the largest quantity downward. This should include water (aqua) but does not need to include sodium or potassium hydroxide as these no longer exist in your final product. Under ingredients, you also need to list allergens that exist within the essential oils if they exceed 0.1 percent of the total weight of a wash-off product, or 0.01 percent of a leave-on product. Ask your essential oil supplier for allergen breakdowns.

You are required to state an average weight for your product on the label and this should be followed by the "e" symbol which must be at least 3 mm high. To ascertain an accurate weight, the law requires that you check your product on "stamped for trade" scales. If your soap bar weighs between 3.5 oz and 7 oz (100 g and 200 g) you have a weight tolerance of 4.5 percent over 40 bars. Since soap loses moisture over time, it is a good idea to label your product at 10 percent less than it actually weighs when you first cut the bars.

Your label also needs to state contact details for your company, your product batch number, date of manufacture, and, if it is a cream or a liquid, the open jar symbol with a "use by" date. If there is any possibility that your soap could be mistaken for food, you also need to put "DO NOT EAT" in bold, large caps on the label.

CLAIMS

You should take great care to not make any claims on your product label that could be construed as "medicinal" as this will throw you into complex additional legislation. For example, you cannot say that a product will "help eczema" unless you have proven this with extensive clinical trials. While you can state that your product contains oils with proven therapeutic properties (for example, "includes shea butter for its moisturizing properties"), you should not say that the soap itself is "moisturizing." If you are manufacturing soap, you must have both public liability and product liability insurance.

SCALING UP

Soap is to a large extent a numbers game. As your profit on each bar is relatively small, you need to sell a lot of soap to make a decent living. When running a business, time is at a premium, and it takes just as long to make a 44 lb (20 kg) batch of soap as it does to make a 2.1 lb (1 kg) batch. The first thing you need to do when scaling up is to create a situation where you can make a single, large batch of basic soap and then divide it into separate containers before adding colors, fragrances, and additives. This means you may only need to actually make soap one or two days a week.

WORKSHOP

Whether you are working in your kitchen or in a purpose-designed space, make sure you have a wet area for the messy stuff and a dry area for packaging. Also remember that soap is slippery, and if you work on polished floors, fresh soap and essential oils can strip the varnish off. Talk to a plumber about fitting a grease trap under your sink. Fresh soap is fatty, and in cold weather fat can solidify and block the pipes. Use paper to wipe off as much soap as possible from your pots and buckets before washing them using dishwashing liquid.

EQUIPMENT

Replace your small buckets for large buckets and your stick blender with an electric drill with a whisk attachment. Keep your mold size manageable, bearing in mind whatever cutting device you intend to use. There are some creative cutting solutions on the market that depend to some extent on the size of your initial block of soap. I've provided a link to a UK cutting machine specialist in the resources section on page 143. Make sure that you choose a mold shape and size that minimizes any waste.

CREATIVE MARKETING

Like it or not, your packaging, brand positioning, name, and niche will make or break your business. Here are a few things to think about:

- It is your image that will position you in the market. Some of the most expensive soaps are sold with minimal packaging.
- There is a lot of competition out there. Focus on a niche market and build a small range that specifically targets it.
- The shape of your soap bar is important. People like a bar that is easy to handle.
- Keep your name short, generic, memorable, and easy to fit on bar soaps and bottles. It also needs to be easy to spell for website users. Do the same with your logo.

FIND YOUR NICHE

Look at all age ranges and focus on where you want to go—all age groups and income brackets use soap. Are you aiming at earth mothers or diva shoppers? The upwardly mobile or the already got there? The tourist or the stuck at home with the kids? Picture your customer and list where you think they buy their clothes, what restaurants they go to, and what newspaper they read. Your unique selling point could also be based on a rare ingredient you include in your product. Some of the most successful brands have been built on just one main product with a special purpose or identified market, think Coca Cola. Think also of the savings you would make buying in just one set of ingredients in bulk rather than a hundred small bottles that you have paid a premium price for and that will sit on your shelves eating away at your profit.

Do not be tempted to produce numerous varieties of soap—keep your proposition simple and cost effective. The same soap formula can be repackaged and resized to serve several markets.

A flexible packaging solution like paper banding will enable you to personalize soaps with your computer printer. Over 60 percent of soaps packaged for the gift market do not actually get used—they just sit there for years looking pretty. There are two markets—soap consumers and soap givers. Soap consumers use more; soap givers pay more—cater for both. And remember that male grooming is a buoyant and growing market.

PACKAGING

At the bottom of each of the bar soap recipes, I have made suggestions for creative packaging, but if you are designing a range to sell through shops, you need a packaging solution that can be handled without damage, projects your image, and tells your story. While you are perfectly capable of wowing a store buyer with your products once they get placed in a shop, you only have your product to talk to the customer for you.

If your range includes soaps, creams, and lotions, make sure that they all sit together well and that each product is easily identifiable as part of the one brand. Packaging is often more costly than the soap itself, so you also need to consider the labor costs. A beautiful, hand-wrapped bar of soap can take up to 10 minutes to achieve, so work that out in wages.

DIRECT SELLING

A huge percentage of the population buys soap of one kind or other, and yet many soapmakers still feel the best outlet for them is a craft market, where there are often other soapmakers competing for trade and lots of people who just look rather than buy. Markets are great because they give you an opportunity to see a customer's reaction to your products, but the best markets are the most unlikely ones. At the risk of sounding sexist, consider typical male events like car rallies and imagine how delighted a bored partner will be when she sees a soap stall oasis she can spend money at.

Here are some more places you could try to sell to:
- Guest houses and hotels
- Florist and gift shops
- Health food stores
- Shopping center (mall) foyers
- Party planning events
- Wedding organizers
- Promotional companies
- In-staff lunchrooms of large corporations
- Internet sites
- Consumer shows where cosmetics are not the focus
- Corporations for their washrooms
- Anywhere where there are lots of people

PR, ADVERTISING, AND PROMOTION

You should budget to spend at least 5 percent of your annual turnover on promoting your company, but, while repeated advertising can work, your budget would be better spent on establishing your image via your name and logo and a well-designed website and/or brochure. Before creating these, do lots of research into what other soapmakers are doing and then avoid doing the same. An original shape, material, or color combination will help your promotional material stand out from the rest.

The best form of promotion is PR, which requires the following:
- The ability to write a great press release.
- Good photographs of your products.
- Creating a good network of publicity contacts.
- Sending soap to key people—your product is your cheapest PR resource.
- A great deal of research and hard work.

WEBSITE MARKETING

Soap is not an ideal product to sell on a website as people like to pick it up and sniff it. It is also heavy and therefore costly to mail. If you are starting a website from scratch, you also need to understand that unless you invest a great deal of money promoting the site, it usually takes at least six months to build a decent hit rate. If you want to sell via the web, link up with established websites where your product will fit.

INDEX

FURTHER READING

The Handmade Soap Book
Melinda Coss, New Holland
ISBN 1-85974-006-5

Gourmet Soaps Made Easy
Melinda Coss, New Holland
ISBN 1-85974-626-8

Scientific Soapmaking
Kevin M. Dunn, Clavicula Press
ISBN 978-1-93565-209-0

Making Natural Liquid Soaps
Catherine Failor, Storey Books
ISBN 978-1-58017-243-1

Soap Naturally
Patrizia Garzena and Marina
Tadiello
Programmer Publishing
ISBN 978-0-97567-640-0

The Aromantic Guide
Kolbjorn Borseth, Aromantic Ltd
ISBN 978-0-95543-231-6

The Fragrant Pharmacy
Ann Worwood, Bantam Books
ISBN 978-0-55340-397-8

Je cree mes Savons au naturel
Leanne and Sylvain Chevalier,
Terre Vivante Editions
ISBN 978-2-36098-007-9

Natural Spa Products
Kolbjorn Borseth, Aromantic Ltd
ISBN 978-0-95543-234-7

RESOURCES

UK

www.soapmakersstore.com
Great range of soap and cosmetic basics.

www.inoviainternational.co.uk
Good when you want to buy in larger quantities.

www.gracefruit.co.uk
Particularly good for those hard-to-find ingredients.

www.akamuti.co.uk
Specialist African oils, butters, and clays.

www.sheabuttercottage.co.uk
More African supplies.

www.sheamooti.co.uk
Distributors of Afrikids' shea butter from Ghana.

www.essentiallyoils.com
Knowledgable and reliable essential oil supplier.

www.aromantic.com
Great for your creams and lotion ingredients.

www.makesoap.biz (author's website)
Soap and cream making courses in the UK and
France. Professional soapmaking molds.

www.laurelseedoil.com
Wonderful bay laurel oil for Savon d'Alep.

www.dormex.co.uk
Excellent jar and bottle supplier.

www.rockinglambs.co.uk/soap.htm
Great soap cutters.

www.cosmeticsafetyassessment.com
Scott Grainger is a highly qualified chemist and
soapmaker, and he is the person to talk to in Europe if
you need cosmetic safety assessments for your recipes.

US

www.snowdriftfarm.com
Great ingredients and lots of recipes and know how.
Also has an online calculator.

www.thesage.com
Another excellent US supplier with a calculator.

www.moldcreations.com
Suppliers of fancy molds.

CANADA

www.oshun.ca
Ingredients on a large scale.

FRANCE

www.aroma-zone.com
Everything you need for soap and creams,
including thermometers and mini scales.

AUSTRALIA

www.soapnaturally.org
Supplier links and lots of resources, plus a soap
makers forum.

www.newdirections.com.au
Stockist of soap and conmetic making ingredients.

SOUTH AFRICA

www.hexachem.co.za
South African arm of Inovia (see under UK).

INTERNET FORUMS

Captain Ethel's Soapmaking and Toiletries Resources
http://freespace.virgin.net/michele.jack/index.html
Lots of brilliant links for useful suppliers
Europe and US.

The following groups are accessed
via **www.yahoogroups.com**

UKsoaping
Very helpful group of UK and European soapmakers
sharing knowledge.

Liquidsoapers
US group for when the bug really gets you.

Soap_makers
US group. Lots of knowledge and lots of traffic.

ACKNOWLEDGMENTS

Many thanks to Clare Sayer and Emma Pattison at New Holland for knocking my idea into shape, applause for Jon Meade and his great photography, and a special thank-you to Carol Benham for correcting my awful 'rithmetic.

A big hug for all my fellow soapmakers and clients who have shared years of ideas and support, in particular Ged Harrison who taught me the cold method and demystified the thickening of liquid soap. Thanks also to Monica at Visionary Soap for letting me relive the dream—more power to you !

Finally a big cheer for the team at Makesoap.biz—Juliette, Chris, Maureen, Jaqui, Sharon and Agnes for enabling me to continue doing what I love to do best, teaching people how to make and sell soap.